The R.A.M.S. Library of Alchemy

Volume 26

The Mineral Work
by
Johan Isaac Hollandus
(Johannes Isaaci Hollandi)

R.A.M.S. Publishing Company

The Mineral Work

by

Johan Isaac Hollandus

(Johannes Isaaci Hollandi)

Produced by

Restorers of Alchemical Manuscripts Society
1980

R.A.M.S. Publishing Company

R.A.M.S. Publishing Company
117 Rutherford Lane
Stuarts Draft VA 24477

The Mineral Work
Copyright © 2015 R.A.M.S. Publishing Company

First Edition 2015

ISBN-13 **978-1511576772**
ISBN-10 **1511576774**

Image Processing by Philip N. Wheeler

This book is sold for informational purposes only. Neither the publisher nor the editor shall be held accountable for the use or misuse of the information in this book.

Printed in the United States of America

Table of Contents

Disclaimer 7

Introduction 9

Hermetic Philosophy 11

PREFACE OF THE TRANSLATOR 17

The Mineral Work 18

A Word from the Publisher 309

Dedicated to Hans W. Nintzel,

American Alchemist

and

Founder of the

Restorers of Alchemical Manuscripts Society

(R.A.M.S.)

Disclaimer

Liability: The publisher does not warrant or assume any legal liability or responsibility for the accuracy, completeness, or usefulness of any information, apparatus, product, or process disclosed. The publisher makes no representation as to the accuracy or completeness of the contents of this book and specifically disclaims any implied warranty of merchantability or fitness for a particular purpose. No warranty may be created or extended by written sales materials or sales representatives. You should obtain professional consultation where appropriate. The publisher shall not be liable for any loss of profit or other commercial or personal damages, including but not limited to special, incidental, consequential, or other damages.

Introduction

Philip N. Wheeler

Johan Isaac Hollandus (late sixteenth Century) is said to have been the two Isaacs Hollandus, father and son, Dutch adepts, who wrote 'De Triplici Ordinari Exiliris et Lapidis Theoria', 'Mineralia Opera Sue de Lapide Philosophico' and other works on Alchemy. The details of their operations on metals may be the most explicit that have been given in writing, and may have been dismissed by some because of this very clarity. John Read, a Professor of Chemistry, in his 'Prelude to Chemistry, an Outline of Alchemy,' dismisses the writing of the Hollandus pair in a few words, possibly because their clarity of detail led him to suspect a ruse.

Hans Nintzel selected this work for inclusion in the R.A.M.S. Library.

THE MINERAL WORK
JOHANNES ISAACI HOLLANDI

In which are contained the drawings of his secret furnaces and several other vessels and instruments mentioned in others of his writings; in addition to other excellent secret techniques.

Translated from Low German into High German by a skilled fancier of

Hermetic Philosophy

PREFACE BY THE TRANSLATOR

Dear reader, after the third part of the Mineral Work of the great philosopher Johannes Isaac Hollandus, of which nothing has heretofore been mentioned in any book, came into my hands, I immediately decided to translate it into our High German language for your benefit. Many a man might not have done it, but might have kept it for himself. I, however, do not consider this right, bit believe it to be everyone's duty to further the common good to the best of one's ability, which furtherance is done in no small measure by making available to people the writings of the wise men, so that they, when reading than, are induced similarly to strive for wisdom, art and virtue, and to appeal for those to the giver of those gifts.

If, then, Hollandus is considered one of the wisest in natural sciences who ever lived in Christendom, it is right that his writings should also be diligently brought to light and not withheld, as is unfortunately no doubt done by some; because one cannot learn anything of his great *Opere vegetabili*--which is no doubt a magnificent intricate book, because everywhere reference is made to it and over 300 chapters of it are quoted — but not of his *Opere animali*. One can only assume, therefore, that these and others of his writings, of which one knows nothing yet, are suppressed and withheld by envious persons; but how right such a procedure is every fancier can see by the fact that when he hears of a good and useful book, he wishes and would like the owner to let him read it too.

Well, then! If you like others to treat you in this way, treat them likewise with what you have, according to the express commandment of our Lord: What you wish others to do unto you, do unto them. Thus I hear that a man of high rank is withholding the Vegetable Book of the dearest man, Paracelsus, nor does he allow anyone to copy anything from it. Those must be devilish minds which, so to speak, rob the public good. Listen, you envious monster, no matter who you are, you who have the audacity to withhold something written for the general public by such highly gifted men. Have then these same authors written for the purpose that you should conceal and hide it? And not rather that the general public and the homeland, also its dear children, the descendants, might be improved by it? If their intention had been to keep their gifts secret and hidden, they might wall have saved themselves the trouble of writing.

But I am afraid that this reflection will be little heeded by such envious persons; for whoever does not take note of the terrible threat of Christ (which will come true still much more assuredly than we see heaven and earth before our eyes) that the servant who buries his talent will be thrown into uttermost darkness, will take much less note of our unimportant words. We will therefore leave them to their poisonous minds and take care of ourselves, that we may do our duty with the help of Cad, by whose impulse I, as mentioned, have most diligently translated this treatise from Dutch into High German, have copied all figures most exactly, and sent everything to be printed.

And I hope, if it pleases God, next also to publish Hollandus' "*Opus vegetabile*" of the wine. Although it has been published in Latin in Arnheim, I have discovered, by comparing it with the Dutch manuscript, that it has not been translated correctly. Besides, it is not complete, but the second part is still missing, not to speak of the fact that one can hardly get it any longer, which deficiencies we hope to compensate for.

Concerning our Hollandus, I cannot really know at what time he lived, but I have heard from a noble *chemico* that he lived at the time of the latter's grandfather with whom, as he was also an excellent philosopher, he was linked in great friendship, just as with yet another great possessor of the secrets. Those three great masters were very famous at that time. This, and other circumstances, allow us to assume that Hollandus had no doubt known Paracelsus, but that he was probably rather old at the time of Paracelsus' youth (1493-1541). The same is probably true of Basilius Valentinus, that he also still lived when Paracelsus was young, although the high-minded Helmont states in the 'Tractat tria prima Chymicorum principia' that Paracelsus was 150 years younger than Basilius Valentinus, and that his invention of the three Chemical Principles had been a plagiarism. It is, however, easy to prove by Basilius Valentinus himself that that opinion is wrong, since Basilius teaches a remedium in his *Trimphant Chariot* for the French disease, which then just occurred. But that disease started only about the 15th century, and neither Basilius nor Paracelsus could therefore have written twenty years apart from each other. And assuming that Basilius and Hollandus wrote twenty years prior to Paracelsus, we can nevertheless not assume that their writings were immediately published, so that Paracelsus could have copied from them and interlarded his writings with them.

It is also true that they have their knowledge from others than Paracelsus, and none of them can be said to be the inventor of the three Chemical Principles and others with which they deal; for it can be proved by very old books that the supposed third *principium* of the alcaline ash-salt which is in all things in the fire, has been known and used by *chymici* since time immemorial, though the schools did not classify it among the columns of the principles, as they are today being taught in the books of our guild. Yet it may well be that Paracelsus consulted orally with these highly-illuminated men and that they guided him to greater scientific knowledge, since Paracelsus traveled in many countries looking for the Art, and was not ashamed to learn from some men. He could also have agreed with them in his writings without their instruction, and be equally talented like them. Just so other *Adepti* have written consonantly in various nations not known to each other, especially since Paracelsus did not lack in any way in incomparable ingeniousness as his *praeceptor, Salomon Trismosin*, recognized and predicted in his school years that his disciple, Philipp Hohenheimer would yet turn into a wonder of the world, as did in fact happen.

Let that be as it may, no one can have a great gift from anywhere but from God, when we should also thank for it and pray to him with all our heart that he might further endow bright—shining men, such as would be highly necessary for the crumbling and horribly confused Church matters, for restoring obedience and faith to order and harmony, so that the miserable yelling of the ravening wolves; here, there, over there is Christ - according to our religion you will be saved - would come to an end; and instead the age-old holy apostolic life would again be set in motion by wise men sent by God; for by their splendid divine gifts and calls to unity we know that teachers come from God.

In contradistinction, dispersion, confusion, disorder, manyheadedness, and countless sects are a sign of the ravening-wolf kind. One should justly learn to notice and understand that God is not a God of confusion and disorder but of order and peace, and that he recommended and left to us peace on the occasion of his last farewell. It is surely a pity that one listens to such miserable wretches of sectarians, allowing oneself to become so deplorably separated from the bond of perfection, that is, from love and unity; since it is publicly known that such lawlers do not understand small earthly things, how then can they have recognized the heavenly things never seen, as our Lord likewise argued: If you do not believe when I speak to you of earthly things, how would you believe if I speak to you of heavenly things? As if he wished to say that, whoever is blind for earthly things which he sees everyday, must necessarily be blinder and more ignorant for heavenly things which no human eye has ever seen.

This incontestably applies to the sectarians, in whom not one spark of divine power, wisdom and virtue is shining, but who only resorts to their sheep's coats. If they were so eager to preach, they would probably do so before unbelieving heathens, which they do not do, however. May the merciful God graciously deliver his Church from such vermin; with these words enough of this.

For the rest I request and admonish anyone to whom it applies, that if Divine Providence has put one or another edifying writing under his care, he make such likewise available for the common good, to which it belongs. Why would he needlessly expose himself to suffer the judgment of the man who hid his talent, while instead, by surrendering his talent to the exchange-bank, he can make himself agreeable to God and pious persons, I, for my part, will earnestly endeavor to recognize such honesty as a favor bestowed upon me *in specie* and a Grace of God to be earned to the best of one's ability. With this, may the Christian reader fare well, and may he be sincerely commended to Divine Grace.

PREFACE OF THE TRANSLATOR

Let it be known to the reader that the arrangement of the chapters in this tractate has not been done by *Hollandus* himself but by me, for better retention. Otherwise, however, I have kept to his style word by word. Apart from this, let it be said to the thoughtless as a sincere warning that they should not undertake the processes herein described and work without reflection according to the letter; otherwise, I will not accept any guilt if they burn their hands. It is better first to labor with the head and to understand the meaning of the Sages through studying rather than through the hand and expenses. The Sages have another *Mercuris* than the common, which is like the other but very different in quality. This must be obtained through the old Saturnus, which carries the sign of the world.

The Mineral Work

CHAPTER I

My child, the philosophers followed nature and first put all things in water, without any *feces*, before they used them in the Chymical Art. Similarly, the philosophers also looked for the earth in the water, just as God Almighty first put the earth into the water; and they did find the earth in the water, which they then called their precious and dear stone, since the beginning of all things is of earth. That is why one must take one's earth, prepare it and make it fertile, before sowing into it; for without preparation it cannot bear fruit, because by itself it is cold and dry. If, therefore, it is not moistened by rain and warmed by the sun, it cannot produce fruit. On the other hand, if it is burnt by the sun, it is also sterile; that is why it must be of the right temperature, not too hot, not too cold, not too dry, not too moist. To this end, the masters invented an *Aquafort*, with which to make the earth hot and moist. In this they dissolved their earth, not pouring on it more strong water (Aquafort) than required. The philosophers write about it as follows:

If our stone is too dry, it brings forth no fruit; if it is too humid, it drowns the fruit; if it is too hot, it evaporates. That is why the earth or the stone must not get more Aquafort than is necessary to dissolve it. Otherwise it will become either too hot, or too

Now they took a large glass vessel that was very thick, and they put the amalgamate into it. They put a helm on with a big head, as big as a man's head, in addition to three or four noses. To every nose they luted a big recipient, and heated moderately for 3 days. After this, they gave a stronger fire for 3 days. Then they made the vessel burning hot for 12 hours. Thus they sublimated all that could be sublimated, but they took care that the vessel was quite tight, because of the *lac virginis*, which goes over each time and drips in the recipients. They kept this *lac virginis* well stoppered till they put the stone into putrefaction. After that, they allowed it to cool down, removed the helm and the sublimated matter. They put the *feces* into a stone mortar, added the sublimate, together with salt and vinegar, till it again became an amalgamate. That they washed with common water. But if it did not come alive, they added a little of the mercury and put the *amalgamation* to sublimate as before. This they did so long till everything rose together through the sublimation. In the manner just related the *corpus* was overpowered and rid of its crudeness, and made spiritual.

CHAPTER IV

The above-mentioned is the first ordinance of the philosophers, and it is an augmentation of the *corpus*. If now you wish to make an *augment* of this white or red chalk (or lime), put it in such a glass (No. 1 of plate pg. 15) and that into a furnace. Revolve the glass, and again, so as to fix it. (figieren). Then you have gold or silver, according to your work.

CHAPTER V

If, however, you wish to make the philosopher's stone, take this fixed *calcem* which has been coagulated in this way, and *imbibire* (saturate or imbibe) it with the aforementioned *lac virginis*. Take it to a furnace and put it into a vessel with ashes or sand. Then take

✳ coagulated with *alcali*, that is, 2 parts of *alcali* and one part of *sal ammoniac*. Put them together on your furnace to sublimate. Repeat this sublimation till the

✳ stays fixed on the bottom. During the night, let it dissolve on a stone in cold air; during the day, in a cold cellar that is humid. Drench the silverlime with this water. Repeat it seven times, drying it each time in a "Cupel" with ashes. Then dissolve it *in balneo* or horse dung, for 40 days, in an open vessel. Now take it out, pour off what has been dissolved, and put what stays at the bottom on a furnace as before. Again drench it with *sal ammoniac* water. Do that 7 times, as before, and let it again dissolve as before. Repeat until everything is dissolved.

Now coagulate it and turn it into a *subtile* (subtle) powder. Put it into a broad vessel, thick as a thumb; let calcinate *in tripode* for 21 days; then take it out and set to *putrefactio*, or into the *balneum*, for 40 days. In between those 40 days you must nourish your medicine with good food of gold or silver. Once you have added it, close your vessel and set it to putrefy; and feed it till it is satiated, for the medicine has become so subtle that it would consume itself and come to naught if it were not nourished, the red with gold, the white with silver. When the pieces no longer dissolve, close the vessel and let it stand thus for another 7 days. Then open your vessel, and throw yet another little piece into it to see if it gets still dissolved. Look for this every 6 or 7 days till the 40 days are over; then take it out and set it to coagulate. Now your philosophical stone is prepared, all metals transmuted into gold or silver.

CHAPTER VI
HOW TO PREPARE SILVER FOR THE STONE, AND ITS FOOD

My child, let us now consider what thing *Hermes* and his descendants found of which they made their perfect stone. They took fine gold and fine silver, as it comes from the *Minera*, and as nature had cooked it, since the seed must be good, if the fruit is to be good. You cannot take any kind of seed for which a fruit is to grow for itself. For whatever seed you sow, such kind you will reap. Do not, therefore, look in a thing for what is not in it, as said before.

And they cemented their gold and silver, i.e., gold (was cemented) 7 times through *cementum regale*; they re-

fined silver on the "Cupel" and refined it of lead. Then
they beat it thin like pennies and cemented it with
common salt which had been cleansed of its earthiness by
having been dissolved, clarified and congealed. With this
salt they cemented the silver so often till it came out
white of the fire. But you must know that you should not
make it too hot, so that the salt can melt; for then the
silver would also melt, and then all this labor would be
lost.

After this, keep it for 24 hours in the following
cement: Take 3 lbs. of white, purified and prepared salt,
as before; Roman vitriol clarified of its earthiness, 1
lb; Living sulphur, of which the crude has been separated
by boiling it in vinegar, as will be taught later on, 1
lb. Dissolve these three together in good, distilled wine
vinegar and put them for 21 days in the *balneum* to
putrefy. After this, congeal them and powder them down
finely; now calcinate them for 24 hours without melting,
following which remove the fire and let them cool down.
Again rub them to a fine powder and let them dissolve in
cold air at night, and in a cold cellar during the day.
When all is dissolved that can be dissolved, put your
matter into a glass vessel; place that above the fire in
a bowl with ashes; add a helm. Thus your matter will stay
pure. Now take as much of this matter as you wish, powder
it finely and, together with the laminated silver, put
layer upon layer in your vessel. Stopper it well and
cement it for 24 hours, as has been said before, without
flowing. Do this as often till your silver has started to
become dead (taub is actually "deaf") and that it meets

24

your wishes. Then your silver is ready for immediately making of it the stone, and also to feed with it before and during *putrefactio*.

CHAPTER VII
THE PURIFICATION OF SULPHUR

Take 10 or 12 lbs of living sulphur, powder and boil it in good, clear wine vinegar till the vinegar is colored. Pour this colored vinegar off, and pour other vinegar on top of the first; again boil it till the vinegar is colored. Pour that one off too and add fresh vinegar. Repeat till the vinegar is no longer colored. Now congeal this colored vinegar, and your purified sulphur; which you must use for the work described above, will stay at the bottom.

ANOTHER PURIFICATION OF SULPHUR

There were some who took living sulphur, 12 lbs, powdered it and put it into a big, earthenware can. They poured distilled vinegar upon it and put the can in the *balneum*, well stoppered. They let it boil for 3 days and 3 nights, then cooled it down and let it settle. They decanted the clear, colored vinegar, and poured fresh vinegar on top, and after the mixture had been boiling for 3 days and 3 nights, they cooled and skimmed it. They repeated this till the vinegar would not become colored. Then they threw away the *feces*, put the colored vinegar all together into an alembic with a helm, and distilled the vinegar to about one quarter in the *balneum*. Then they took it out and poured it into a glass dish. They placed it for 3 or 4 days into a cold cellar, and within that time it turned into a saltpetre, clear and transparent like clear "Augstein" or a noble golden yellow. Again they evaporated that vinegar in the dish till something stayed in it, and they put it again into the cellar to crystallize. They evaporated to a dry, yellow powder what was not crystallized, and which was so beautiful as the powder of noble gold. They also let the little stones evaporate into a yellow powder, and again dissolved it in distilled wine vinegar, as before. They then precipitated the *feces*, and poured the pure matter from above, and again fresh vinegar on top. They repeated this work till no more feces remained.

Again they distilled the colored (tincted) vinegar, approximately three-quarters, and let it crystalize as before. Since all the sulphur was by now clean and pure, without leaving any feces, it was ready and cleansed, pure as crystal which is saved. It is a great alchemical secret how to purify sulphur in this way.

CHAPTER VIII
THE PURIFICATION OF MERCURIUS

Take Roman vitriol, 6 or 8 lbs; common salt, 2 lbs. Mix them with 3 lbs of mercury (quicksilver), which has previously been well washed with salt and vinegar. Sublimate your mercury three times through them, each time taking fresh matter. Keep it for later use.

CHAPTER IX

THE PURIFICATION OF SAL AMMONIAC

Take sal ammoniac, 3 lbs; sublimate it through Roman vitriol and *Lap haematit*, or bloodstone, 3 times, each time with fresh matter; and keep it for when you need it.

CHAPTER X

DISSOLVING WATER OF THE PHILOSOPHERS

They took Roman vitriol, 6 parts; *Lap haematit* or blood-stone, *crocus ferri* or iron rust, *vermillion* or cinnabar, *AESUSTUM* or burnt copper, *ammonium minerale ana*, one part. They dried these items till they powdered, then put them into a distilling vessel, poured rectified *aqua vitae* on them, 4 lbs, and distilled them again to the finely powdered *feces*, three times one after another. They divided this water into two parts, each part into a separate separation glass.

Now they added one-third of the prepared ✳ to the red, sublimated and preserved as mentioned before, and dissolved in the *balneum*. When they had done that, they dissolved in the same glass one-third of the prepared ♁, also in the *balneum*, and later also the ☿. There those

27

three were dissolved in the water of the philosophers, which had been prepared from the *aqua vitae*, and which is rightly called the philosophers' water on account of its wonderful secret powers. Its wonders cannot be described, nor is it fitting to describe them, because of certain reasons. They preserved this water in which the spirits had thus been dissolved. Then they dissolved one-third of fine gold in the other part of the water, in *balneum*, until all gold—varnish was dissolved into clear water.

After this, they poured it to the water in which the spirits had been dissolved, stirred and mixed it well in order to unite the spirit with the body. Then they sealed the glass *hermetice* (hermetically), put it into a dish with strained ashes on the stove, as much ashes as the matter was deep. At first, they gave but a little fire, and then they saw the spirits go up and down together with the water, with many little veins which were blood red and golden yellow, until the spirits turned into one color, i.e., brownish-yellow. Now they increased the fire somewhat and kept it thus steadily till they saw the little veins become bigger and coarser and fewer. That was a sign that the matter began to thicken and started coagulating. Now they increased their fire once again, and the coarser and thicker the little veins or rays became, the stronger and stronger they made their fire, until the matter was fixed and no more rays could be seen in the glass. The matter stayed fixed at the bottom, like oil, brownish-red, clear and transparent. This oil was the elixir, *compositum*.

Then they took this elixir, put it into an *ampulla*, sealed it, put it *in tripode*, and gave a moderate heat for 40 days and nights. Within this time the elixir coagulated into the philosophical stone, so that one could turn it into powder. It was a delightful treasure,

which changed lead into gold, just as copper, ♀ and ☽, which gold was better than the one that comes out of the mountains. Remember this work well, the *projection* will teach itself.

CHAPTER XI

ANOTHER WORK WITH THESE THREE SPIRITS

My child, they then went further and accomplished yet another work with these three spirits. They took ✳ , ♀ and ♁, prepared as before, of each one-third ℥ , mixed them. Then they took one-third of iron—fillings, one ounce of copper fillings, and one-third ℥ of grated lead or white lead. They mixed these pieces and dissolved them in the normal way in *aqua-fort*,

made of 6 parts of Roman vitriol, 3 parts of ♂, 1 part of *lap haematit*, and 10 parts of saltpetre. This they poured 3 or 4 times on the feces, each time powdered finely and so dried that they were like dust.

29

They divided this *aq. fort* into two parts. In one part they dissolved the three spirits; in the other part they dissolved the *corpora* (bodies), such as iron, copper and lead. When they were all dissolved into clear water, they poured both waters together into a glass, and put it, well closed, in the *balneum*, to putrefy for 15 days. Afterwards they drive the aq. fort off in a moderate balneum, till it dripped no longer. After that they left it still another 8 days in the balneum, boiling till the matter was dry. Now they removed the glass from the balneum and put the matter into another glass, as is depicted on the plate as No. 2. they sealed it above and put it in tripode to <u>digest</u> there for 15 days, and to dry well and calcinate.

Now they took the glass out and put the matter into a sublimating vessel. They sublimated it, first with a little fire; afterwords stronger. They did this 7 times, each again mixing what had risen with the feces, in order to thoroughly draw the spirits out of the bodies. Then they put the feces to reverberate into the reverberation furnace, during 3 days and 3 nights, with a moderate heat, letting them glow blood-red, but not bright-red, for that would be too hot.

After that, they took it out and put the matter into a glass, poured distilled wine-vinegar on it, and placed it for 3 or 4 days into the balneum. Every day they stirred it 3 or 4 times with the hands; then they let it cool and let the feces drop. They poured off what was pure very gently from above, put a helm on, and distilled the vinegar off. Thus the salt of the three bodies stayed at the bottom. They removed this salt and calcinated it again in the reverberation-furnace and let it well glow for 7 days and 7 nights. Then they took it out again, put it into a glass, poured good distilled wine vinegar upon it, put it in the balneum, and did as before. Then they took out the salt which was very clear and blood-red.

Then they took our Aqua vitae prepared for the Red, as it is taught, and poured it upon this salt, and dissolved it on hot ashes. Again they drew off the Salt in a lukewarm balneo; they poured fresh Aqua vitae on it and dissolved again, and drew it off in the balneo as before. They repeated this till the salt no longer congealed but remained like a red oil. Then they took the sublimated spirits, ground them on a stone, and imbibed them with the dissolved salt, steadily rubbing and drying (them) at the sun or on lukewarm ashes, till all the oil had been absorbed by the spirits. Now they took all the matter and put it into a glass, such as is drawn above, and put it in tripode, giving a moderate fire for 15 days. After that, they removed it, broke the glass, for the matter was as hard as glass; then they pounded the matter to a subtle powder and put it into a fixation glass ("Figier—Glas"). They poured some of our Aqua vitae upon it, which dissolved immediately. But they poured some more on it till everything was dissolved. Then they sealed the coagulation glass *hermetice* and put it on the furnace. They let it stand on warm ashes, gradually increasing the fire as it was going up and down till everything was fixed and remained at the bottom as a fixed oil which is an elixir. They put it into a glass *ampulla*, sealed it above, and kept the glass for 40 days in tripode, regulating the fire as before. Within this time the elixir congeals into the philosopher's stone which truly transmutes ☿, ♀ and ☽ into ☉.

CHAPTER XII

Now I will teach my child how to make the furnace of the philosophers, which is their calcination—furnace or a dry "Stove", for in this furnace they calcinate their sublimated spirits and congeal therein their elixir. They also digest therein with a gentle fire, that is, they make their matter subtle and dry their *salia* in it. The name of this furnace is *tripus*. My child should know that we have heretofore given instructions on many kinds of works, without separation of the elements, for which this secret furnace is often used. Now, however, I will teach you some works in the course of which this secret furnace comes in handy several times. Make it as follows.

MODEL OF THE SECRET FURNACE

No. 3

First one puts down a big, round foot of the furnace, in
which there are 3 or 4 ash—holes. Starting from these
ash—holes, one puts masonry 2 feet high. Upon that one
puts an iron cross that is big, strong and thick (No. 4.)
One continues to make a

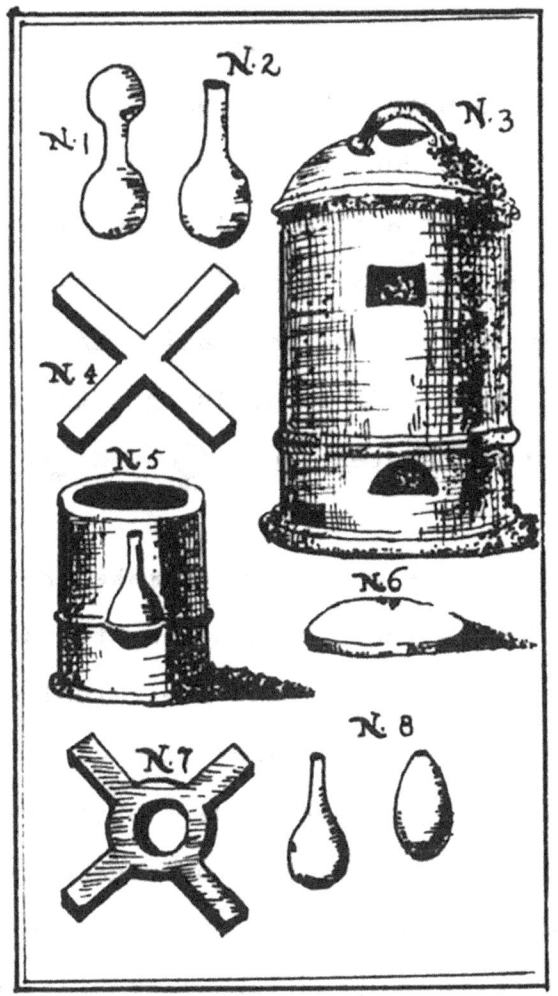

round furnace with
masonry, one and a half
ell high. In the center
of the furnace, one
leaves a square hole,
into which one can
insert one's hand to
feel the heat of the
fire, together with a
stone which closes it.
After removing one's
hand, one must
immediately close the
hole to preserve the
heat. Let the furnace
be coated inside and
outside with well—
keeping glue. Over the
brim of the furnace
leave a groove, the width of two fingers, since there
must come on top a big, raised lid, inside well glazed by

the potter, and outside well plastered up with glue. Inside the furnace a chamber (room) must be constructed (No. 5 of the diagram), one ell high, baked of strong clay in a potter's oven. It must be 4 fingers thick, well glazed inside as well as the one end with which it will stand on the iron cross which lies at the bottom of the furnace. This vessel should be wide, one and a half quarter, so that one quarter of room remains between the container and the big round furnace. An even lid (No. 6) must be made for the inside, and in the center there should lie a cross with a round ring (No.7), upon which stands the glass *ampulla* (No. 8) or the philosopher's egg. When either is standing on the cross, one should cover the chamber (container) with its lid and lute it.

After this, one has to tightly lute the large, raised lid onto the inner rim of the big furnace. Then prod the fire below in the furnace. In the chamber, stove, or dry oven there stands the glass with the matter, or the spirits are calcinated therein, or one can dry therein.

My child, understand me well as far as this earthenware vessel is concerned that is to stand in the big furnace. It is called the chamber or dry stove. The potter is to make it, and inside the iron cross must lie, somewhat lower than in the center. On this cross is placed the *ampulla* or the egg, or another open vessel in which one wants to dry something. I have also drawn three types of glasses that we need in the Art.

34

In this furnace you can calcinate all spirits
without their volatilizing, for in this furnace they
cannot fly, because they have everywhere equal heat, most
of all in the upper part of the furnace. For the
earthenware cask is standing on the big cross, one and a
half foot above the fire, and it stands a quarter ell
away from the furnace on all sides, so that the heat
flows around the earthenware vessel between the walls of
the furnace, and rises against the lid. Then the heat
turns down again and upon the earthenware cask; and the
earthenware cask is up above luted with its lid, so that
no air can enter it. Thus the spirits have equal heat
round about in this furnace. Therefore, one can dry all
spirits in it and produce their crystallization. In this
furnace one achieves that all spirits and *corpora* unite
and merge into one another. You calcinate in this furnace
all spirits, to dissolve and distill them afterwards. If
we did not possess this furnace, we could do nothing. In
this furnace all elixirs, or fixed oils, convert into the
philosopher's stone. And know, my child, that if you did
not have this furnace, you could not work in the Art,
even if you had the art of all philosophers. That is why
all philosophers keep their furnaces secret, as the
utmost secret. For without this furnace one cannot
accomplish the stone; that is why they have called this
furnace the secret furnace of the philosophers and *tripus*
or *Fimus Calidus* or *Stercora*, and with many other names
too long to tell here.

CHAPTER XIII

THE CEMENTING FURNACE

No. 9

The cementing furnace must be round and thick
above, and closed all around. Above, in the round hood,
there must be four or five holes, wide enough to allow a
man to put his little finger into it, since air is
passing through there. In the center of the furnace there
should be a wheel with many holes, lying on a star that
reflects the flame. On the wheel there should be three
teeth on which stands the crucible, so that the flames
cannot touch it. At the side of the furnace, above the
wheel, there should be a square hole to insert the
crucible. It must again be closed with a stopper that
fits into it, and glued, luted during cementation.

CHAPTER XIV

Aqua Fort AND *Aqua Regis* FOR THE
SOLUTION OF GOLD AND SILVER

Now we will again turn to the prepared gold and
silver. After either has been prepared, one has to have
various *Aqua Fortis* to dissolve the gold and silver. My
child, in order to make an *Aqua Fort* for dissolving gold,

take saltpetre, vitriol Romanum, common salt and ✳. To
dissolve silver, take saltpetre and *alumen richae ana*.
However, before making this water, all materials must be
prepared ahead of time, that is, first one has to

dissolve, clarify and congeal them, and let them shoot forth, after which they must be well dried. When they are well prepared and dry, put them into an earthenware vessel which must be built in such a way that its mouth would fit into a Syburgian ruffle, in which one should receive the water. This earthenware vessel must be luted one thumb thick, and with the same lute the necks must also be glued together. Such a lute is made as follows:

The white of eggs, fresh cheese, vintager's butts of buttermilk, good rye flour, *bolus*, and soaked paper. Everything is to be well mixed together and your vessels are to be luted with that mixture. Let them well dry, first give A GENTLE FIRE LIKE HOT SUNSHINE, FOR 24 HOURS, JUST AS IF ONE WERE TO KEEP LEAD IN FLUX WITHOUT DRIVING IT. Afterwards, increase your fire so that the pot becomes gradually glowing hot, for another 24 hours. Let it stand in the same heat, irrespective if no water goes over, 12 hours. For there are still fixed spirits going over, which improve the work wonderfully. Then let it cool down and preserve this water. *Nota*, in the receiver there must be clear Aqua Fort, at the rate of 2 ounces to 1 lb of the matter, so that the spirits of the matter can all the better move into the ∇ ; thereafter, take once again prepared matter, according to how big your work is. Put it into a glass pot with a helm which has a big head and 2 or 3 beaks, (No. 10) large enough to allow one to put a thumb through them. Above, in the top of the furnace there should also be a hole, through which the Aqua Fort can be poured upon the matter. After careful

37

luting, give a gentle fire, when the receivers are attached, for 24 hours. Now increase your fire gradually for 24 hours, till it becomes burning hot again, as mentioned previously. Let it cool down, and add the Aqua Fort to the first. Now take again new matter, distill it as before. You must do that 9 times. The glass, however, into which all this water is to be poured, must be quite large.

Now take the *capita mortua* (the dead heads), let them dissolve, clarify, congeal and shoot up, as I have taught you elsewhere. When they are quite clean, take as much saltpetre as the *capita mortua* weigh; dissolve them together and congeal them so that they are well dry. Then put them into a big glass pot and pour all your distilled Aqua Fort upon it, and give fire as for the first 7 days, and again for 3 days as if one wished to keep lead in flux without glow. After this, another 12 hours in glow (strong fire), then let it stand for 3 days in order to cool down. Now you have the philosophers' water, with which one can do many wondrous things. Its power cannot be comprehended, for it turns all bodies into spirits and all spirits into bodies. It destroys everything put into it; it is a work of wonder in our Art. With this water the Masters shorten the work of one year to one month, and of one month to one day.

CHAPTER XV

My child should now take his gold, or silver, prepared as I instructed before; it must be laminated and cut. You must put it into a big recipient and pour on it as much Aqua Fort as to enable it to dissolve into clear water. Place it on a furnace in hot ashes or sand, so that it may well dissolve in such a way that no white clouds remain in it when it is cooled down; but it must be dissolved purely, without one's noticing any feces in it. Now set it in the balnuem in order to separate the wateriness from it, and the White from the Red; and give it something to eat of the aforementioned gold or silver till it is no longer hungry. Let it stand such as, not hotter than to allow you to dip your right hand into it. When it is no longer hungry, distill the water off it in balneo, to the thickness of a child's gruel or somewhat thicker. If you draw off the *phlegma* or the water in such a way, look after the helm, for if it threatens to turn red or yellow, distill no more, but let it cool down.

When it is cold, distill again till the helm turns red or yellow; then let it cool down again. Do this work till you have drawn off all wateriness, or till you no longer see the sign. Remove the helm, and give it again something to eat, as before, and feed it well with small pieces, as before. Close the alembic with a cork and luting; then put it on the furnace in ashes or sand. As often as you close it, you must lute above with prepared lute of wheat flour, egg-white, fresh cheese, butts of buttermilk, *Bolus*, etc.

My child must know that if your medicine is dissolved and stays dissolved for a long time, it becomes so spiritual that one could never again bring it into a

39

corpus and melt it. Therefore, it must be nourished, otherwise it would fly away and you would lose your work, especially since this Aqua Fort is very fierce when its phlegma are gone. It is then called the Red Lion, so fiery it is and so hot a stomach it has; also a dragon that devours everything. That is why it has to be fed and satiated with good food, that is, with prepared gold or silver, beaten very thin. It has to be constantly under observation because of the feeding, which is the noblest part of the work: to throw small pieces into it, so that it should not eat more. The vessel must each time be luted above. Put it on the furnace in hot ashes. Finally, make it somewhat hotter, as the Aqua Fort becomes gradually weaker, because it has swallowed much gold and silver.

You must also know that the alembic must be opened and again closed twice a day, so as to give air. When it no longer demands to eat, let it nevertheless stand on hot ashes for 2 or 3 days. After this, pour your medicine from the alembic into a double Venetian glass (No. 11)*, because the medicine has to stay in this glass till it is perfect to do *projection* with it. This vessel's size has to be according to the scope of your work. Now pour it into a big alembic or recipient of Hessian glass (No. 12). Cut that crosswise with a red hot iron wire in order to lift off the upper part, as necessity demands. Put your glass with the medicine in the lower part, on a golden, or silver, or glass tripod. Lute caps on it with the aforementioned luting; also mix some quicklime with it, which keeps against water. One has to remove and put on this cap often; but you must most diligently lute, otherwise it would spoil and would not keep.

Now put the helm on the alembic and place the alembic on the furnace in ashes or sand. Give it fire as if you would burn rosewater, so that the red spirits do not rise from the Aqua Fort of the medicine. The red spirits must stay with it and be coagulated with it, and the medicine must be distilled as dry as gruel or children's pap, but not more dry because of certain reasons. If it were dryer, it would not open up during *putrefaction*; if it were too moist, the *solution* would give too much water and would dissolve too soon, which would not be good for the medicine. It would then not become subtle enough.

CHAPTER XVI

My child must know that there is a great difference between putrefying and keeping in the balneum, for in the balneo one distills, but in putrefying one does not distill. In addition, in the putrefaction the medicine is well mixed and merged and rendered subtle, so as to congeal afterwards and then to coagulate, as you will be taught later. But know that you must lute your helm and recipient tightly, and thus prepared put in ashes; and distill with little fire, according to the above instruction, the wateriness from the medicine after it has been fed. Take care, however, that the red spirits do not also rise. For this reason do in everything as you have been taught. When your medicine is distilled and is

like a pap or gruel, remove the helm and put again a
stopper on the alembic. Lute it well above and put it
into putrefaction for 40 days in boiling hot water, and
always take care that it be boiling hot, without
interruption, or your matter would spoil. Nota, the
little glass with the medicine must also be closed with
its cap and well luted, when it is put into putrefaction.
(see No. 14)

CHAPTER XVII

THE CONSTRUCTION OF THE PUTREFACTION FURNACE

Make a round stove, of one stone's thickness (No.
15 page 34) high 2 ells. Hang a kettle therein which is
deep 1 ell; or a deep pot, which hangs in the furnace on
2 or 3 books, fastened inside the furnace. There should
only be room for 2 or 3 fingers' width between the
furnace and the kettle, so that the heat can play all
around the kettle to the same degree. In this kettle
there should be a tripod, 4 fingers high. Upon the tripod
yet another kettle has to be placed, three-quarters high;
and between it and the other kettle there should likewise
be room for 2 or 3 fingers. This space is to be filled
with small hay. Then water is to be poured on it up to
the uppermost brim. It must always be kept filled that
much, and every day and every night must be filled once
with warm water. You must also fill

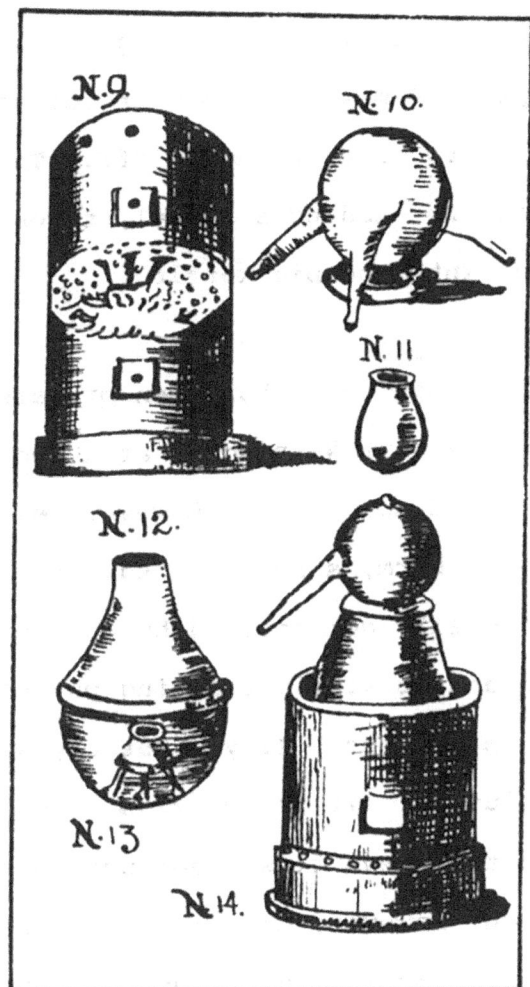

the inner kettle with hay at the bottom. Into that hay put your alembic in which is your medicine. Stuff the inner kettle full of hay around the alembic, so that it stands firm and the kettle is quite full; however, the hay in the inner kettle must stay quite dry. That is why it must be covered and well glued with a leaden or copper plate, to prevent any smoke from entering it from the hot water.

Afterwards, put on the furnace a bell (dome) that fits it, made of potter's clay, well glazed outside and inside, having a hole above, three fingers width, to allow the steam to escape. It must not have any other hole. It should have two handles to lift it on and off. There should also be a hole at the side for adding a funnel, through which the warm water is daily poured into the big kettle.

CHAPTER XVIII

After all things have thus been prepared, give fire for 40 days and nights, always keeping it at a temperature close to boiling, not hotter and not colder; however, even if it were boiling, there would be no harm, since the medicine is so well preserved that it cannot easily get spoiled by boiling water. Nevertheless, you must watch and be vigilant during putrefaction, for it is the easiest thing in the Art. If you putrefy well, you will also produce a good solution. It is impossible to join some spirits or bodies, unless they have first been put into pure water, without feces. Such, however, cannot be done perfectly, except by means of the putrefaction and solution. Therefore, putrefy your corpora well, and also your spirits, and join them thus dissolved, then you may reach perfection.

My child must know that putrefaction renders the work so subtle that it is not possible to change it back into a corpus, solely through digesting, of which we shall give instruction later. Know also that it should stay in putrefaction until everything has turned into clear water, without leaving any feces at the bottom. Then it is duly putrefied and, when it is thus well putrefied, you can draw off the *lac virginis* and bring the work to a powder for emergencies, ("notherheischliches"), as will taught later.

CHAPTER XIX

When the medicine has been well dissolved, without leaving any feces, open the alembic, put a helm on it, and put it on the aforementioned distilling furnace to coagulate till it is dry. Draw off the *lac virginis*, which lac virginis is the element air, yellow like an oil. After this, draw the element fire likewise off, and the feces (earth) will stay at the bottom, black as coal. Now take out your medicine. It will be between hard and soft, more soft than hard; it will disappear in contact with cold air, and congeal over fire. Such must be the case. Besides, during putrefaction it has absorbed moisture. That is why it is dissolved in air. This moisture must be removed through correct *digestion*.

CHAPTER XX
SETTING UP THE DIGESTIVE FURNACE (No. 16)

First construct a round wall, of the thickness of two stones and the height of 1 ell. Inside the walls, half an ell and two fingers width from the earth there should be a hole to put in there the fire or the coal. Construct another furnace on this foot (with masonry), a quarter of an ell high and of the thickness of one stone. Approximately half an ell up construct a square hole, into which one can put the hand to feel how one should regulate the fire. One must put a stone, a cover, or lid into it to go in and out as one wishes. The furnace must be round inside, from the bottom to the top, and well glued inside with *luto sapientiae*, so that it is not affected by the fire.

Upon this furnace there should be an iron ring with 4 hooks, on which should hang, in the furnace, a thick metal or copper vessel, the thicker the better, on account of the heat. This vessel should be one ell high and, if possible, two fingers' width thick. Round about it there should remain two fingers' width space between the walls, so that the heat can play all around the kettle and rise and heat the whole furnace. Put on this kettle a lid of the same material that can close the kettle, since it must get well luted on the kettle to prevent any air from escaping. Dry, sifted ashes have to be put into this kettle.

You should also have an earthenware vessel, made of clay one thumb thick, large, wide and high enough to allow the alembic with the small glass containing the medicine to stand in it. Between this vessel and the kettle there must likewise be two fingers' width of space.

Put this cask in the dry ashes; yet the earthenware vessel must not be glazed over, just as the alembic and its head (Haube) must not be covered when it stands in the earthenware cask. The earthenware cask, however, must have a well-fitting lid, which must not be luted on but only lie tightly fitting on it. And know that this earthenware cask and the ashes must be quite dry before you put your medicine into it.

To put the medicine in, do as follows: First, take the alembic with your medicine; put it into the earthenware cask and cover the latter with its lid - or put it uncovered into the copper vessel or kettle, upon whose bottom there is two fingers' thickness of ashes. Then hang this kettle in the furnace, cover the earthenware cask with its lid, and fill the kettle all around the earthenware cask up to the latter's lid with dry ashes; and cover the kettle with its lid; lute it so that it gets no air. Then cover also the furnace, like the Putrefaction Furnace, because it is alike, except for the hole into which one inserts the funnel which is not required here. These two furnaces are certainly not identical, for in the Putrefaction Furnace there were two kettle, whereas here there is one kettle and one cask, although this kettle is bigger than the other. Yet with cleverness one could also manage with one furnace. Now lute the lid so that no air can penetrate through it except through the hole that is above. When everything is ready, give fire as is required.

CHAPTER XXI

Digesting means attracting the superfluous moisture that is in some medicines, either its own or that which it has absorbed during putrefaction. This excessive moisture must be consumed by dry heat, more or less according to what is required. Since a thing which melts in the air and coagulates in the fire has excessive moisture in it, and when that is gone, it will melt in

47

the fire and stay (unchanged) in the air, as common salt,
ammoniac, tartar, *sal alcali*; those are fusible by nature. Even so, they dissolve in the air and congeal in
the fire. That happens because of the excessive moisture
they contain. But when that (the excessive moisture) has
been removed from them by proper digestion, they become
fusible and can be dissolved by fire and harden in the
air.

According to how much excessive moisture things have,
one has to give them fire. One requires more fire than
another. Things that have a great deal of moisture must
in *digestione* be given fire as if one wanted to sublimate
mercury (quicksilver), constant, equal heat, as best as
you are able to. Continue with it till your matter is no
longer dissolved in the air, then it is enough. Test it
in this way: Open your kettle and also the earthenware
cask, and with a spoon take out some of your matter from
the glass. Put it on a stone in a humid cellar. If it
dissolves, it is not sufficiently digested. Consequently,
you must put it again to digest by giving fire as before.
Instead, if it is not dissolved, it is enough.

The aforementioned method holds good for the
digestion of all matters. But you must well take care
that your matter does not become all too dry during
digestion, since it would then not be fusible. That is
why it is best to keep to the middle between moist and
dry; and if it must be, it is better too moist than too
dry. Take for example borax, whose powder will not
dissolve in the air, no matter how it (the air) be. Never-

theless, put it powdered upon a stone in a humid cellar -
it will get dissolved, although *Borax* is between hot and
dry and moist. That is why it retains its fusibility,
because the humidity of the cellar affects a subtle
powder more than a hard piece that is not powdered. Borax
contains moisture in itself, because it is sprouted in
sugar—water, just as alum and vitriol are in pure water.
Nevertheless, borax is easy to melt, because it is not
too dry. Something that is too dry is difficult to melt,
as one may clearly see from alum, vitriol, common salt,
etc. They do not easily melt on account of their great
dryness. That is why I say that a fusible matter had better be too moist rather than too dry. One can see it with
common salt, ammoniac, sal alcali and tartar, which are
fusible by nature. But when they are purified of their
earthiness, they are much more fusible than before, the
sole reason being that they now have more moisture than
before. That is why they would be naturally fusible, even
if their moisture were so much diminished that they would
sprout in water.

Understand: Everything that is too dry does not
easily melt; and everything that is too moist, is easily
melted; one can remove from it its excessive moisture.
But to restore its moisture to that which is too dry
would take too long. It is as if one tried to make glass
fusible on account of its great dryness. A thing that
sprouts in good water - one has to evaporate its
wateriness till a little skin shows. When you see that
sign, put it in a cold, dry place; if it sprouts, it is
moderately (or: medium) fusible.

Thus you may now, relying on the above—mentioned criteria, test all things whether they are easily meltable or not, whether they are too moist or too dry; and if something in your work was not to your liking, you can get rid of its deficiencies, since the work depends on this. In this way you must well explore this matter, etc.

CHAPTER XXII

Now we will continue to make our powder naturally fusible through fire during digestion. When your matter stands in digestion as mentioned heretofore, you must give as much heat as will allow you to hold your hand, after inserting it through the hole, without burning, between the furnace and the kettle for the period of one Ave Maria. At this temperature the fire must be kept for 15 days, from time to time testing with your hand through the hole. After that, the hole must always be well stoppered, because the medicine has no other moisture than from the Aqua Fort and from putrefaction. That is why the medicine does not require a big digestion. If it were too dry and you would immediately put it back into putrefaction, it would nevertheless not get dissolved.

CHAPTER XXIII

How TO ADD THE SPIRITS AND TO PREPARE THE ELIXIR THEREOF, WHICH IS THE MOST PRECIOUS TREASURE IN THIS WORLD

When the 15 days are over, remove your medicine or
matter from *digestion*, and thank God. Then one can do
projection with it, and it is now ready to achieve the
great elixir. It is also easy to join with it any kind of
mineral spirits or *qu. essent.* and preserve.

These spirits must previously have stood in
digestione, in order to be processed along, as has been
mentioned before. For this purpose, however, the fire
must not be hotter than sunshine at the end of the month
of August. When then the spirits, such as *lac virginis*,
salmiac, *aqua ardens* or other moist spirits have thus
been congealed *in digestione*, they are prepared.

My child must know that one cannot set any spirits in
any kind of work to coagulate, or to remain with the
corpora, or to turn into medicine, unless that to which
one wishes to join the spirits first medicine and so
strong that it could *transfer* all imperfect bodies or
metals into gold or silver. When that is the case,
however, the spirits are protected so that they do not
fly away when they come into the fire. The spirits cause
a deeper entrance of the medicine into the metals; and
they defend that which they meet and resembles them; but
they devour with the fire that which is not like them. In
this way the medicine united with the spirits has more
power than of its own; for as the medicine, when it was
small, had the power to rise during projection, it has
later, after being coagulated with the spirits, the power
to transfer 100 times. And if to the two were also added
the *qu. essentia* of gold or silver, and they were

coagulated together, it would have the power to transfer a thousand times into true gold or silver. And if one's business were to subtilize the *medicine* thus *conjoined* with the spirits and the *qu. essentia*, which would have to be done by dissolution, putrefaction and congealing, it would gain an infinite *projection* in every *subtilisation*; yea, as could not be expressed.

My child should also know that no corpus, except gold or silver, can keep its spirit or quint essentia with it in the fire, unless the corpus be first made likewise (or: evenly) subtle and a medicine, as previously mentioned; that it is able, therefore, to make a projection.

CHAPTER XXIV
How TO MAKE THE GREAT ELIXIR FROM THE AFOREMENTIONED MEDICINE

Let my child take his medicine, made of gold or silver, well digested, as has just been taught. Upon it pour some of your aqua fort, no more than that the work become like a paste or gruel. Then add the head (Haube) onto the little glass and arrange it in the big recipient. Lute its lid on as well, as shown above. But above all other things, set it first into putrefaction, and lute the head onto the little glass in which the matter is. When everything has been prepared, give fire, as has been taught above regarding putrefaction, not to subtilize but solely to dissolve. When the medicine has thus been prepared into its clear water (which may be done within 24 days or less), and it is pure without feces, take your glass and have at hand the above-mentioned sublimated *Mercurius*, or which is sublimated in the following manner:

Take *sublimatum*, as much as there was of your medicine before you put it into putrefaction. Rub it very small on a stone; add it to this dissolved medicine, immediately close your little glass with its lid and lute it, so that no spirits fly away. Then put it again into putrefaction and give fire, as has been taught before — for 40 days and nights. Then look at it: If your medicine has been dissolved into pure water, it is all right. If not, put it back into putrefaction, until it is completely dissolved. Then remove the lid from the little glass that contains the medicine, and again lute the lid tightly on the recipient, put a helm on it, set it in ashes, and congeal your medicine with such warmth as you made for putrefaction, as dry as you can. Remove the helm and close the recipient above with a little glass. Put it into *digestione* in order to coagulate, and give fire as taught before, during 21 days.

Within 6 days, all colors of the world can be seen and finally come to one color. Now remove your glass from the recipient and put it into the earthenware cask for *digestion* as before; give fire for 15 days, as above, so as to decrease the moisture. Then take it out and thank God, for you have the big elixir, which transmutes all metals into gold or silver. But because the aqua fort is with it, one cannot use it in medicaments; otherwise it is better in projection than mineral gold or silver.

CHAPTER XXV

HOW NOW TO JOIN TO IT THE Q. Ess. OF GOLD OR SILVER

Take the medicine that you have just made and which is joined to ☿; pour upon it as much aqua fort as before; put it into putrefaction, and when everything has turned into water as before, take Qu. Ess. of gold or silver, or of *antimonium* (which Qu. Ess. I will teach you to make here), the weight of your sublimated ☿, or your medicine when you first put it into putrefaction. Add it to your dissolved medicine, which has been dissolved in putrefaction. Then quickly close the little glass and afterwards the big recipient, and put it to putrefy like above. Give fire for 40 days as before, and afterwards put it to congeal, as before, and thereafter for 15 days into *digestion*, as mentioned above. Then you will get your desire, with the help of God who bestows such upon you.

With this, do projection one to one thousand or more, upon all metals whichever you wish. The work for the Red or the White is all one labor; but the one requires that the spirit be sublimated to the Red, while the other to the White. The Qu. Ess. to the Red is from gold or ♂, the White from ☽ and *arsenicum*. For the rest

there is no difference in the labor, either for the Red
or the White.

CHAPTER XXVI
THE PREPARATION OF ALL THE THINGS
WE NEED FOR THE SAID WORK

First, one has to prepare common salt as follows: Dissolve salt in common water; congeal the pure. Then take
legs of horses or sheep which are well burnt, 3 parts and
purified salt, 1 part. Together, those are 4 parts. Let
them calcinate for 6 weeks in heat, well mixed together.
Yet the fire must not be too strong, so that the salt
does not burn away into glass. Put it in the kind of heat
that one gives to "Cupel" when refining silver. Take
care, however, that the flames do not touch the vessel.
After this, dissolve your salt in water and separate it
from the ashes of the legs; congeal it and be very
careful that no dust fall into it. Now put it into an
earthenware pot for 3 hours, without melting; again pour
distilled vinegar on it, and let it gradually dissolve;
when it is dissolved, pour off the pure and put it to
distill. When the vinegar has gone down dry, take it out
and preserve it. With this salt you must cement your gold
or silver which is to be added to your work; or you may
make the medicine with this salt and with gold or silver,
in the following manner:

Take Aqua fort, made of saltpetre and alum. With that dissolve of your salt ♌ ii cemented so that it be just dissolved. Then take silver, cemented 7 times with the abovementioned salt. Dissolve that also in Aqua fort, made of ☿ and saltpetre, that it gets dissolved rightly. Then pour these two waters together into a *phiole*. Put this on hot ashes and feed your medicine well, as you have been taught, till it is well satiated. Then draw off the *phlegma* in *balneo*. After that, put it to putrefy for 15 days, and congeal it again, after which keep it again for 15 days in digestione. Now you can make projection on *laton*, which has a soul in it, and on tin. The work is the same for the Red, except that there must be ammoniac in the Aqua fort.

CHAPTER XXVII
HOW TO PREPARE THE MATTER WITH
WHICH GOLD IS CEMENTED

Take new, unbaked, raw tiles (or: bricks), which have never been in the fire. Pound and rub these unfeelingly. Then put them into a potter's furnace to calcinate till they are white, which causes them to become very dry. In the same way the salt with which one calcinates must also be quite dry.

Note: If you cement your gold or silver with "Minifarben" (may be a lead oxide), and therefore you do not draw off their humidity with dry things, they will never be able to receive the tincture of the Aqua fort. Therefore, cement carefully, at least 7 times; if you are after perfection.

CHAPTER XXVIII

Sublimation OF THE QUICKSILVER FOR THE RED

Quicksilver is sublimated to the Red in the following manner: Take Aqua fort, made of Roman ♁, alum *de roche*, *alumine plumoso* and *ammoniac*. In that you dissolve your quicksilver. Again draw off the water through the helm in balneo.

Then your ☿ is dead.

Then take Roman ♁ 1 lb, saltpetre 2 lbs, sal ammoniac 1 lb, alumen de roche 1 lb, common salt 3 lbs. Pulverize all these things and mix them well. Now divide the powder into 3 parts. Then take 1 lb of your dead quicksilver; mix it with 1 part of the said powder and put it into a glass; put an alembicum on it with two noses the size of a man's head. The spouts must be very wide. Add a recipient to every spout, well luted, and put each of them into a dish with cold water. Give the first a mild fire for 12 hours, and stronger for another 12 hours. After that, let your pot burn well for another 12 hours, in which time your quicksilver will rise into the helm as white as snow. It will have within itself the red Qu. Ess. out of which things have been sublimated. The water will be in the recipients; keep it tightly closed in a glass.

Now remove the sublimated ☿ from the helm; mix it again with the second third of the said powder, and proceed with it as with the previous. Do the same with the third part of the powder, and preserve your water each time well stoppered.

With that you will get your quicksilver snowwhite and full of tincture. Powder that finely on a stone and add it to the water which you have preserved from it. Put it on the furnace in a dish with ashes; put a helm on and lute the recipients on to it. Now sublimate as before. Repeat this work 3 or 4 times, till no more feces stay at the bottom, but everything be sublimated together. Preserve this drawn off water well, because you must dissolve your quicksilver in it, and it is better than any kind of water you could make. When your quicksilver has been sublimated and prepared in this way, it is 1 lb of gold and is well prepared for the work of adding it to the medicine of the gold.

CHAPTER XXIX

HOW TO DRAW OUT ALL TINCTURES

All tinctures, white as well as red, are all drawn out in the same way. This work for the Red and the White is a great secret. My child must take and fill a large well glazed vessel with old, clear urine. Put a big helm and one recipient on it, and distill everything that can

it above with a cork, shake it about and incorporate it well, and again put it in the sand or ashes. First give a gentle fire, so that it becomes moderately warm. Meanwhile, remove the cork in order to give air. Otherwise, the glass would burst. Stir it occasionally to mix it well, as before, so that the vinegar can work in it. When you notice that your vinegar and your urine are well colored, pour the pure liquid off above and take care that no feces are poured out. Keep what you have poured off well stoppered and by itself. Now pour again as much urine on the feces as before, as well as distilled vinegar; stopper it and put it on the furnace; warm and mix it as before, and when it is colored again, pour it off and to the other, and preserve it well stoppered. Again pour fresh urine and some vinegar upon the feces, as before; put it on the furnace, heat it and mix it; then pour off the pure, and repeat till all tincture is extracted. Now throw the feces away, or use them for whatever you need them.

Pour the tincture poured together into an alembic, put a helm on with a recipient luted thereon; distill the humidity in ashes or sand. Then the tincture will stay at the bottom red or white, according to the matter used to prepare it. This is the Qu. Ess. of the thing from which it has been drawn. In this way the Qu. Ess. of quicksilver is extracted, sublimated for the Red or the White. *Item*, from *crocus martis*, from iron files, *verdigris, aes ustum*, cinnabar, gold and silver lime (calx) and from *antimonium*; briefly, from all mineral matters you wish. But to make the effect of the urine strong, you can each time add ammoniac and common salt which have been prepared, of each one-third, and extract the tinctures, as has been taught here. In these extracted tinctures you may now prepare cements and cement therewith, which is an excellent art. You may make Aqua fort with it, as red as blood, glistening like a ruby, with which water one can perform many a wonderful thing, which it would serve no purpose to reveal all.

CHAPTER XXXI

SUBLIMATIO MERCURII TO THE WHITE
(THE SUBLIMATION OF MERCURY TO THE WHITE)

Aqua fort, made of cadmia (or: calamine) and "Eierkalk", of each 1 lb; alum de roche and white one: each 1 lb; arsenicum 1 lb; saltpetre 5 lbs. Powder all these things and mix them well together, and distill your Aqua Fort with "Kraucken", (a crock), as has been taught above. After that, rectify it. When it is rectified, add raw quicksilver which has come from the mineral. Dissolve it and draw the water off in balneo, and you will find your ☿ at the bottom, congealed.

Now take cadmia, egg shells (lime or chalk of eggs), white vitriol, arsenic, *alumen plumosum*, of each 1 lb; alum de roche, prepared common salt, each 2½ lbs; saltpetre 4 lbs. Powder all these very fine, well mixed, and divide this powder into 5 parts. Then take 1 lb of the dissolved ☿ ij, mix it with one of the 5 parts and let it sublimate, as was reported before in the Red work.

Then you will find your ☿ sublimated to the White, when it is sublimated with all 5 parts.

CHAPTER XXXII
HOW TO DRAW THE QUINTESSENCE FROM GOLD, SILVER AND OTHER BODIES

My child, take gold or silver or that upon which your work is aimed. Cement it 7 times, as has been taught. The silver, however, must be calcinated until it comes beautiful and white out of the salt, as is indicated above. Take your gold or silver, prepared as previously according to what your work is to be. Beat it very fine or file it small, and amalgamate it with a good portion of quicksilver. Mix it on a stone with distilled vinegar in which clean, common salt has been dissolved; or with a stone pestle in a stone mortar, till it is well

amalgamate. Mix this amalgamate with ij of sublimated ☿

and 2 parts of ♁. When it is well mixed, put your matter into 2 crucibles which lock one on top of the other. With a punch drill a small hole in the upper part and insert a thin piece of wood. Lute the two crucibles firmly upon each other with 2 fingers' thickness of lute. Let the piece of wood stick in the crucible, and let the lute dry well. When it is dry, remove the piece of wood. In that way the wateriness can seep out of the matter. Now put it into sand upon the furnace and first give a gentle fire till all humidity has gone out of it. Then close the hole on top and let it dry.

Now increase the fire till the crucible is glowing below, but not above. Let it stand thus for one hour. After that, make the crucible glowing hot throughout, but carefully, so that the silver does not melt. Let it stand for a half hour and then cool down, and you will find that the silver has been calcinated into a subtle lime (or: chalk) and a very fine powder ("ungreiflich" or impalpable).

Now take tartar ("Weinstein") that has been calcinated white and sal ammoniac, half a pound of each. Put them together into an alembic, pour upon them a pint of white, distilled wine vinegar which should be quite strong. Add one "Mark" of the calcinated silver, or as much as your work requires, and immediately close your alembic tightly to prevent air from entering it, as otherwise the strength of the ammoniac would escape. Lute your glass tightly and put it into putrefaction for 21 days in equal heat, as has been taught above. After this time, open your vessel, put a helm on, set it in sand to distill. First, the vinegar will go over, afterwards the Qu. Ess. of ☉ or ☽, and will rise into the form of the quicksilver. This quicksilver is very powerful in our work and no less necessary, as you will hear later.

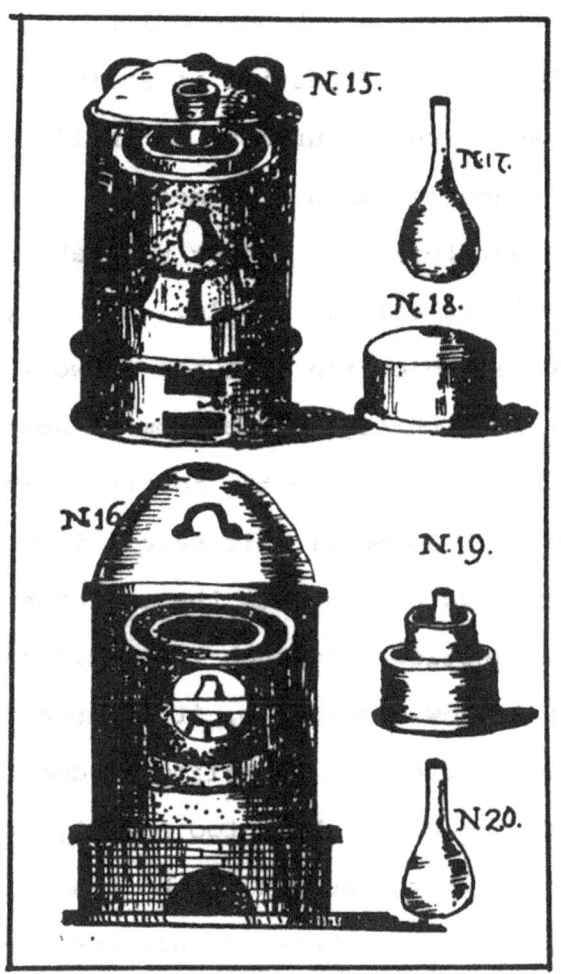

CHAPTER XXXIII

Now I will teach my son how he should finish and complete his work. My child should know that there are 7 orders in our work, which one has all to perform before the work attains to perfection. The first is cementation; the second, dissolution; the third, feeding; the fourth, putrefaction; the fifth, coagulation; the sixth, digestion; the seventh, fixation ("figieren"). You must also know that all these works are the same for the Red or the White, except for the seventh, the fixing; because for the Red it must stand much longer in the fire than for the White. In addition, the Red requires more fire in the fixation than the White, since the Red must be imitating (or: resembling) the nature of a man, but the White like the nature of a woman. That is why the Red must stand longer and hotter than the White, so long till it changes into a beautiful color resembling gold, by which one recognizes that it is perfect.

My child must also know that the wise stone, through cementing, changes from cold and moist into hot and dry; but during solution it will again become moist; in putrefaction it becomes subtle, and in digestion it again becomes dry, coarse, earthy, and fixed. Further, my child must know that the medicine is congealed when it comes out of putrefaction and digestion. Therefore, one cannot yet make projection with it, because it is all too subtle. And if you nevertheless were to make a projection with it, it would not evaporate in the fire, yet flow through the bottom on account of its subtleness. This is so because the fourth element is not with it, since that must also be with it before it becomes perfect. For if one did not add a corpus to the medicine, one could not make a perfect projection, although the medicine were applied; for, although it is now a medicine, it is nevertheless not a corpus; and on account of its subtility it does not stay on an imperfect corpus, because such is too earthly and coarse (or: earthy). Yes, even if you put it on the best silver, it would not produce anything on it, because they are foreign to and different from each other because of the great subtleness. That is why we will now add to the medicine the fourth element.

Remove your medicine from digestion and divide it into 3 equal parts. Rub it on a clean stone; take as much of the white silver calx (which I will here teach you to prepare) as one third of your medicine weighs. Then you have 4 parts, that is, 3 parts of medicine and 1 part of silver calx. Mix well. Now you must get dew, gathered in the morning in a fine cloth. Distill it over a gentle fire; rub your powder with it, or the medicine mixed with silver calx, like soap or butter. Now dry it in a glass dish so that it may powder. Thereupon put it into a round glass like No. 17* and lute, with firm lute, a round glass upon its mouth. Now put your glass into an earthenware vessel No. 18* made like a box or covered dish. The size must be such that round about between the glass and the earthenware vessel there must be 2 fingers' width filled with salt or sand, No. 19*. On top of it yet another vessel should be locked, like a quiver, and the uppermost vessel is to have a little hole to fill through it with sand and salt. Then close it, lute the joints tightly with *luto sapientiae*. When it is full, put a small piece of brick on the small hole and lute it well with *lutum*. Afterwards lute the earthenware vessel all around 2 fingers' thick. When it is dry, set it in tripode on the cross, which is in the furnace. Cover the furnace with its three heavens and give it a gentle fire to start with, for 3 days. After that, increase the fire for 3 more days, as if one wished to keep lead in flux, in a reasonable way. Then your medicine will be joined to the earth or the 4th element. This is now perfect and has the power to transfer all imperfect bodies into its like, after which the imperfect corpus has a soul.

CHAPTER XXXIV

All medicines and spirits must be fixed in the aforementioned manner, in the Little Work as much as in the Great Work. There are, however, two ways of making the fixation, which some masters call digestion. The first concerns dry things, no matter how varied or not varied they may be for conjoining. And know that the regimen of the fire during fixation is not one and the same, but must be established according to what is in the fire; and the same applies to the length of time during which dry things are to be kept in tripode, as has been taught before; and one has to proceed according to what has been reported above for the fixation.

The other manner of fixation concerns humid things, when one wishes to conjoin, or bring into one being, things that are humid. These have to be put into putrefaction, as has been taught before, with such a regulation of the fire, also the length and the order of the works according to what you have been put in (to putrefy). But the vessels and the furnace, and the fixation of humid and dry things must be in accordance with all rules. Therefore of one hundred not one reaches perfection, because they do not keep to the rules of the works, although they have them well laid down in writings.

CHAPTER XXXV
HOW TO MAKE SILVER CALX
(OR SILVER CHALK)

Take the finest silver you can get; laminate it as thin as white pennies, (Weispfennige), and set it to calcinate with common salt, which has been prepared and cleansed of its earthiness by dissolving and congealing. In this salt you must calcinate your silver till it comes out white from the salt. Then powder your silver very fine in an iron mortar, in order afterwards to mix it with the medicine, that is, with dew on the stone, as mentioned before.

Or you can calcinate the silver with ☿ as before, and then wash the calx with salt and vinegar; and afterwards with good distilled water from the balneum, till the water runs off clear. Then, when the calx has dried in a stoppered glass, it is prepared for being added to the medicine.

The work for the Red and the White is in everything the same; gold must be cemented 7 times, also calcinated and pulverized as before. It must likewise be added to the medicine on the furnace, though it must get much more fire. Neither can one put it into fixation like silver, since gold is fixed and much hotter than silver.

CHAPTER XXXVI
THE FIRST QUESTION

My child would like to ask me if the Art is indeed possible? I reply yes, for God has created all things perfect, to remain without deficiencies forever; but because of sin the four elements, which were simple in all things, have become *composita*, corrupted and full of harmful feces. Owing to this, many sicknesses and illnesses have occurred in nature, including death. From this we can conclude that in all things made of the four elements there are two kinds of natures, one within, which is eternal and imperishable, and that is the Qu. Essentia. The other part is imperfect, corruptible; many harmful feces have entered it on account of our sins, which harmful feces have been made of the four elements, which they also contain in themselves. These have led us and all things to death. Whoever could separate the poisonous feces from the elements through the Art, and restore the elements into their first being, or into perfection, the elements would then no longer be called elements but Qu. Essentia. For the above-cited reasons, it is therefore evident that it is certain and possible to use this Art in the work.

CHAPTER XXXVII
THE SECOND QUESTION

Out of what is this Art extracted, or in what is this Art contained? Answer: In all things, in water, in the sea, the mountains, trees, fallow grounds, morasses, earth, secret chambers, birds, fish, animals, excrements, urine, hair, nails, blood, eggs, and in all things that have originated in the four elements.

CHAPTER XXXVIII
THE THIRD QUESTION

How many kinds of stones are there? Answer: Many. The first is called *lapis mineralis*, that is, one can prepare the stone from all minerals or all metals. The second is called *lapis vegetabilis*, that is, that the Vegetable Stone can be pulled out of all growing and green things that sprout out of the earth, such as plants, herbs, fruits, trees, wine, honey, and all greening things. It serves man's health and to *incerate* metals. The third is called *lapis naturalis animalis*, or the Soul Stone, and is made out of all animals, birds, sewers, stables, dungheaps, eggs, blood, bones, nails and the like, and serves *Alchymia*. The fourth is called *compositus* and is drawn from many different matters and through the Art joined into a stone. It serves in Alchymy to *incerite* all leprous metals.

CHAPTER XXXIX

My child would now like to put the question, whether perhaps there are more stones? Thereupon I answer, yes, there are two other stones which God bestows upon us for nothing; for, since they grow, they are not noticed. One is for the Red, the other for the White. Anybody may take of them without asking anyone. That is why *Alchindus* says: You should not open your purse in order to incur expenses in this Art; and they are means between metals

and quicksilver, or ☿ does not like to stay with the metals without means, as little as the soul can stay in man without the spirit, which is an intermediary (or: a mediator) between the soul and the body.

And I will explain to you the nourishment of the child in the womb. The latter is there for 40 days before

71

all its members are made of the four elements; and nature is so noble and faithful that it compares itself with the nature of heaven, because of the noble substance and subtlety it has within itself; and this thing has a spirit. For as fast as the spirit is prepared as it should (since it is subtle and spiritual), the soul falls into it by order of God. Thus the spirit is spread everywhere through the body, for the sake of the nature it has from the body. Thus the soul is kept in the body and cannot separate from it. And when the body gradually increases, the Spirit grows with the body and all members. It is the same with these two stones, be it white or red. These 2 stones are Roman vitriol for the Red, and rock alum (alumen roche) for the White. My child should note that the two blessed Stones are called *Mineren*, that is, *Mercurius sulphureus*. Of these two stones *Morienus* says that there are two stones, white and red, to which God the Lord has given such blessing. (see Glauber -HWN)

CHAPTER XL

THE FIFTH QUESTION

Among all these, which is the noblest, most useful, slowest, and shortest to use? Reply: I say that the Mineral Stone is the slowest and requires the most care ("sorglichste"), and is made with concern and danger, since the Mineral Stone is compact and fixed. It consists of 2 waters, of which the first water renders the stone volatile without labor and danger. One lets it dissolve in it. This water makes *Mercurius vivus* mortal and multiplies all colors of the medicine or the ferment. It must be produced by the alchemist with special understanding, so that the spirits do not fly away. It is drawn out of the stone which God has given us free. This Mineral Stone is drawn out with care and danger and requires a lot of time. And it is very useful.

The *Lapis animalis* has no end, because of the separation of the elements and because there is less knowledge of the art of preparing it; it lasts longer than any other stone. Therefore, there is no greater secret, and it is the greatest art ever discovered in nature to extract this stone; and it transforms anything, as I will later prove.

The third stone is called *Lapis vegetabilis*. That is the greening one, and it has hidden powers, more than the other stone; and it is the greatest and most praised among all stones, and its power is above that of all others, for it has been tested in the rectification of the elements and their preparation. It accomplishes the great *elixir*, which is the black one, doing infinite things, more than the others. All the gold of the alchemists, if it is prepared with corrodent (caustic) waters or other corrosives, cannot be used in medicine on account of the corrosives. The gold, however, that is made from the Qu. Ess. of this Vegetable Stone, is not deficient, if it is fixed ("figiert") with the ferment and is made with intelligence by the alchemist. Such gold is medicinal and not corrosive, nor deficient, since no corrosive is added to the regimen of the Vegetable Stone.

That is the reason why the gold made with it is medicinal and Qu. Ess., and passes in the first degree. This stone is so full of power that, if it is made of silver and thrown upon gold, it will transform gold into silver; instead, if it is made of gold and is thrown upon silver, it (the silver) will turn into gold.

You will find this quality in the Vegetable Stone and not in any other; because this stone, or Qu. Ess., is in no other, neither in the Mineral nor in the Animal, and it becomes the greatest medicine in man's body, preserving it in good health to the last goal of his life.

The fourth stone is called *Lapis compositus*. It is composed of many different spirits and bodies. In its operation one finds great trouble and pains, in the opening of the vessels to prevent the spirits from flying away, The work is tedious and difficult, and it is a great secret of nature and a stone of great power. It is accomplished with great labor, care, danger, and in a long time.

CHAPTER XLI
THE SIXTH QUESTION

My child, you have asked whether in nature there are still more secrets useful to alchemy. Answer: I report that there are many different *opera* (works) in the Art. For not all men know the stone which God has given us free, be it white or red. Nor can all men produce this stone, not the Mineral Stone, nor the Animal Stone, nor the Vegetable Stone, nor the Composite Stone. Thus they do not have much and would like to get advantages out of the Art. For this reason they undertake many kinds of works.

CHAPTER XLII
THE FIRST WORK

There are some who are making odd amalgamations of ☿, ☉ and ☽. They boil them on fire till they stay together at the bottom. After that they dissolve them together in an Aqua fort. They draw the water off one or two parts, then let it crystallize into little stones. After that, they congeal it again and dissolve it once more. They are interchanging this work so often till it stands like oil. Then they take sublimated to the Red or the White, depending on their work, taking 7 parts of ☿ij vivus to 1 part of oil. They mix the oil with the sublimate on a marble. After that, they put it into a small fixation glass, boil it hot, sublimate it and mix it again with the feces, until one is fixed with the other. Then they dissolve it with Aqua vitae distilled and prepared for that purpose. Then they draw off the Aqua vitae in a lukewarm balneum till it turns into a powder. Thereupon they pour fresh Aqua vitae on it and dissolve it again. They are doing this 7 times, till it no longer congeals but stays like an oil. A bit of sugar, like unmelted honey, and it is a perfect elixir which transforms all imperfect metals into true ⊙ and ☽, better than they come out of the mines.

CHAPTER XLIII
THE SECOND WORK

There are some who make a cinnabar from quicksilver that has been well washed with salt and vinegar, thereafter sublimated and rectified in boiling water, and of sulphur, clarified like a crystal. Of this they make a *vermilion*, taking 2 parts of quicksilver and 1 part of sulphur, powdered on a stone and put into a pot; boiled on fire as one makes cinnabar. Afterwards they calcinate the vermilion and dissolve it in *aqua salis* (salt water), prepared for that purpose; they imbibed (drench) 7 parts of cinnabar with 1 part of aqua salis, kept it on a small fire for 7 days, in a closed cask; afterwards, they imbibed it again and did as before. They did this 3 times, until 3 parts of aqua salis had been soaked into the 7 parts of vermilion. Then they set it to be fixed ("figieren"), till everything remained together fixed at the bottom. Then they dissolved it in aqua vitae prepared for that purpose, and put it immediately into a fixation glass; they kept it in warm ashes, rising and falling, till everything stayed fixed together at the bottom, and was a true elixir.

CHAPTER XLIV
THE THIRD WORK

Others took living ☿, washed with salt and vinegar, and sublimated it 7 times through one pound of vitriole, each time with fresh matter. In addition, they dissolved gold in rectified aqua vitae, and rubbed the sublimated ☿ to a subtle powder, which they put into a glass pot and imbibed with the water from the gold; put a helm on, and drew it off the ashes with pure heat, till it was dry. Then they again rubbed the powder quite *subtil* on a stone, put it back into the glass pot together with the helm on it, and drew it off in warm ashes as before. They repeated this work till 3 ℥ of the water were imbibed in 7 ℥ of mercurial powder. Each time, one ounce of water of the gold should be soaked into the above-mentioned powder; and do away with the moisture which you must draw off with the alembic, because it is no good.

When all the water from the gold had been soaked into the powder, they put it to sublimate on a gentle fire, and afterwards on stronger fire. That which rises must again be mixed with the feces. Let it sublimate again, till it no longer sublimates up but stays together fixed at the bottom of the vessel. Then they poured *aes ustum*, *crocus martis* over. In that they dissolved ☿, congealed it on warm ashes, drawing off the moisture with an alembic. Then they put it again to dissolve in balneo, or *fimo*, or in a cold cellar, till it dissolved into pure water. That they congealed again on warm ashes as before, and then dissolved again. They repeated this aften till it no longer coagulated but stayed thick like honey. Then

it is a true elixir, penetrating and tincturing ☿into true gold, which is better and more highly colored than that which comes from the mines.

CHAPTER XLV

THE FOURTH WORK

Some took living ☿, and washed it with salt and vinegar till it was clear as a mirror. Then they took Roman vitriole, 4 parts; aes ustum and crocus martis, 2 parts; *sal commun*, 1 part. Through that they sublimated 7 parts of mercury, first with little fire, then with stronger; and they did this 7 times, each time with fresh matter. After this, they took well rectified aqua vitae and *antimonium mineralia, aeria usti, crocus martis, cinnabar, ana* 2 parts; as much vitriol as the weight of all the others together. They dried the vitriol till it powdered, and then mixed everything together powdered very fine. In a glass, they poured upon it well rectified aqua vitae, rising two hands' width above the matter. They stirred it well together and put it in the balneum, after well luting the glass, for 10 days, Every day they stirred it 3 or 4 times with the hands, so that the aqua vitae might all the better get incorporated with the matter. After the 10th day, they put a helm on the glass and distilled first with a little fire, and finally for

12 hours with a strong fire, so that all its colored
spirits might go over with the aqua vitae.

They repeated this work often, each time pouring the
water upon fresh matter, till the water was as red as
blood. Then they took 1 lb of this water and dissolved in
it as much gold as would dissolve. Thus they obtained a
water of gold. Then they took 1 lb of that water and
dissolved in it as much of the mercury sublimated above
to the Red as could be. After this, they took one part of
the mercury water and two parts of gold water and poured
both waters together into an alembic, well mixed. A helm
upon it, and set it into a dish with ashes, and drew off
the moisture with a gentle fire. Then they took two parts
of mercurial water and one part of gold water and poured
it upon the powder which they had congealed. They put it
back on the furnace, congealed it again as before, with
little fire. They repeated this work till both waters
were united in a *massa* or powder. Then they put it to
solve on a stone, at night, in fresh air, or in a humid
cellar, or in dung, or in balneo, till all was dissolved
into pure water. That which was dissolved was congealed
on warm ashes. They did the solution and congealing so
often till it no longer congealed but stayed like an oil,
fixed, transparent, red as blood, and eternally fixed.

Then it was a perfect elixir which tinctures ☿ into true
gold and goes through all leprous metals and their
diseases ("Malzey").

CHAPTER XLVI

THE FIFTH WORK, TO SUBLIMATE TO

THE RED FROM ☉ AND ☿

Some have calcinated ☉ to a subtle powder, and washed ☿ with salt and vinegar as clear as a mirror. Afterwards they sublimated it through Roman ♁, aes ustum, crocum martis, and ♀ is, hematite ("Blutstein") – of the last four as much as one as of the other; and

♁ as much as the weight of the others together; also some common salt of which there was no weight. They sublimated this 7 times in our sublimating vessel, each time taking fresh matter. Then they took aes ustum, cinnabar, crocus martis and ♀ is hematite, antimonium, ana, Roman vitriol as much as the weight of all the others together, and made a subtle powder of it. They poured on it two hands' width of rectified aqua vitae and put it in a warm balneum for 3 days, stirring the matter daily to mix the matter well with the aqua vitae. Afterwards they distilled the aqua vitae off through the helm, at first with a gentle fire, which they increased gradually, and finally with a strong fire for 12 hours, to well draw out all spirits. They did this 3 times, each time with fresh matter.

Then they took 1 lb of this water and in it dissolved as much of the sublimate as they could. Then they took 3 lbs of gold calx, finely powdered, and imbibed it with ℥ iij *aqua* ☿. They put it in a glass with a helm on, and set it in warm ashes, and drew off the moisture with a gentle fire, and congealed it. Afterwards, they ground it subtle on a marble; and again imbibed with ℥ iij aqua ☿, always grinding (powdering). They repeated this procedure till all aqua ☿ had been imbued into the powder. Then they ground it on a stone, and put it into a sublimation vessel; and, while continually rubbing it (mixing it) with its feces of earth or ferment, till everything stayed together at the bottom fixed. There-after, they dissolved it in the same water in which the sublimate had been dissolved; and they put it into the balneum in a well stoppered glass for 15 hours; after that they congealed it on ashes. They repeated this dissolving and congealing till it no longer hardened but remained in the consistence of a golden oil, which is the true elixir which tinges and changes the ☿ and all imperfect metals into true gold, which is better than that which comes out of the mines, constant in all tests.

CHAPTER XLVII
THE SIXTH WORK

In that way some have calcinated the *corpora* and dissolved the *spirits*, and then imbibed the bodies with the spirits. Others sublimated the spirits, and dissolved the *corpora*; and with the dissolved bodies they imbibed the sublimated *spirits*. Then all was equally well, and at the end they achieved an elixir of the same power.

CHAPTER XLVIII
THE SEVENTH WORK

Finally, some took flowing ("running") ☿, and washed it with salt and vinegar till it was as clear as a mirror. Then they took *alumen roche* and evaporated it on a small fire, while retaining its spirits, as dry until it became powdered. Of this alum they took 3 lbs *lapis calaminaris*, *calx ovorum*, salt, of each 1 lb; washed ☿, 3 lbs. They ground everything together on a stone, and sublimated it first on a small fire and subsequently, on a stronger fire. They did this sublimation 7 times, to enable them to imbibe into the ☿ the white spirits of the alum; since alum is full of white spirits, as the Roman (alum) is full of red spirits; for people whiten copper with *lapis calaminaris*, without any other addition, yes, even as white as silver. Then one leaves eggshells in vinegar, and the *calces* are often imbibed with the dissolved eggshells; then these *calces* are reduced into a corpus as white as silver. This those people well know who can make *albationes* (whitenings) to deceive others by that. Thus they imbibed ☿ full with the white spirits. Then they took *alumen roche*, *arsenicum sublimated* to the white, as I have instructed you with ☿; 2 lbs eggshell—chalk, *lapis calaminaris ana*, 1 lb; as much saltpetre as everything else together, that is, 16 lbs. Of that they took just 4 lbs to make *aqua fort*, as you have been taught; thus they made aqua fort 4 times, and they poured all 4 waters into a glass.

Then they took all the *capita mortua* (dead heads) of the 16 lbs, turned them into the finest powder and put that into their distillating vessel. They poured on it all the water burnt from it, and kept it for 15 days in balneo, in order to unite the spirits. After this, they took it out of the balneum, set a helm on it, and made aqua fort, first with little fire and later with a strong heat, for 24 hours. They repeated this work 4 times, so as to bring the strength of the earth into the water. This is the very best water ever found in the world, to serve the work for the white stone or the white elixir.

Then they took 1 lb of this water, dissolved in it as much fine cupel-silver as they could dissolve in it, and filled it, as is usual (or: customary). They poured off the water, and rubbed the calx with my distilled water till the water was gone. Then, they took 1 lb of the calx and 1 lb of ☿, sublimated to the white. They ground both together to a powder, and sublimated it first with little fire and afterwards with stronger fire. They repeated this sublimation 5 or 6 times, each time mixing the sublimate with its feces. Finally they took the sublimate and preserved it in a sealed glass. They put the feces which had stayed at the bottom into a reverberation furnace and reverberated them for 3 days, hot enough that the earth was glowing and no more, (just enough), that it glowed just a little red; for if it were heated too much, it would turn into glass.

After this, they let it cool down and then ground it on a stone with distilled vinegar, put it into a glass jar (or jug), and poured yet a good quantity of distilled wine vinegar on it. They set it into a fresh, boiling hot balneum for 4 days, shaking it every day 3 or 4 times with their hands. At last they let the balneum cool down and the feces precipitate; they poured the pure liquid off above, dried the feces, and reverberated those again for 24 hours in the reverberation furnace. Then they took them out, ground them with vinegar as before, and dissolved them in vinegar. They repeated this till all the feces were dissolved in the vinegar. Then they poured the vinegar together and set it in the Balneum Mariae,

distilled the vinegar in the balneum. Then the salt of ☽ stayed at the bottom as white as snow.

Now they put the salt once again into the calcination furnace to reverberate for 24 hours, after which they took it out, pounded it on a stone with distilled wine vinegar, and dissolved it as indicated above. After the last distillation of the vinegar, they took out the salt which was shining like a crystal and melting like wax.

Then they took the ☿, in which the silver-spirit had been hidden during sublimation, and they pounded them together into one. Then they put it into a fixation-glass. After this, they took *alumen roche, lapis calaminares, calx ovorum, arsenicium commune ana*, and evaporated the alum over a gentle fire, while preserving the spirits; they powdered it and rubbed it together on a stone very finely; they put the powder into a glass or jar and poured well rectified aqua vitae on it, 3 hands' width above the matter, and set it to distill, first with little fire and later with stronger fire, for 24 hours.

Then they took the *caput mortuum*, pounded it finely on a stone and again added to it that which had been distilled, as above. They did this 4 times, *cohobating* it each time with the feces, so as to gain the strength of the earth. And this was <u>unknown</u> to the old philosophers, and serves for the white stone or elixir; and it is a medium between the living ☿ and the bodies, since the spirit cannot get at the body without a means. They took this water and poured it into a fixation-glass, upon the powder, so that it stood above it by 1 hands' width. They sealed it hermetice and put it into a dish with ashes, at first with a gentle fire, till everything was opened in water, which happened in 21 days. Then they increased the fire, and they saw that by the fixation-glass when streaks went through it; for at first they are subtle, but afterwards they become coarse and thick. Then they increased their fire till it no longer rose but stayed fixed at the bottom, in the consistency of a thick oil. When that was cold, it was thicker than honey and was a perfect elixir, tinging and translating (transforming) all imperfect bodies into true ☽, better than which comes from the mines.

If one wished to bring (turn) this elixir into the philosopher's stone, like a *CRYSTAL*, one would have to take a small glass and set it into the secret furnace, hermetice sealed, or in the dry furnace which the philosophers call *tripus*; give fire so strong that you can easily keep your hand in it. In this heat it has to stand for 40 days, when it will congeal into the philosopher's stone, clear as crystal, so that one can powder it; and its power has increased tenfold in nobility and *projection*.

CHAPTER XLIX
THE EIGHTH WORK

All the details used in this work of ☽ must be used in dealing with ♃, since all philosophers agree that the *calces jovis* effect in all works as much as the chalk of ☽; and ☿ drawn from ♄ accomplishes in the work as much as ☿ drawn óut of ☽.

CHAPTER L
THE NINTH WORK OF THE
Quinta Essentia Lunae

Others took silver calx, made with common aqua fort, beaten down with common water, and dried. They took a large quantity thereof and drew from it living ☿, in the following way.

They took good triple distilled vinegar, which they poured into a stone jar. Into that they poured *calcini tartarum* (calx of tartar) and clear, transparent ✳ *ana* of the aforementioned vinegar, 6 quarts, *calces lunae iij*; they took this silver chalk, tartar and ground them 3 times together on a marble slate with the distilled vinegar, so finely that a painter could paint with it with a brush. Then they put it into a large jar of 12 quarts, and poured upon it 1 quart of distilled vinegar. They quickly closed the jar, as otherwise the vinegar would immediately fly out of the jar. They shook the jar with their hands so that the matter would well get mixed with the vinegar. Then they put this jar for 24 days to putrefy in a warm balneum, stirring the matter every day 3 or 4 times, to mix it well.

After 21 days, they removed the can, put a helm on it, set it on a furnace and distilled the vinegar off. Afterwards, the *qu. ess. Lunae* followed in the form of a living quicksilver. It has such a great virtue that it is not permitted to reveal all its power, for certain reasons. And this is the greatest secret in the Art, for this ☿ or qu. ess. Lunae is so noble and pure that it can actually compare with the nature of heaven in its wonderworks, as will be taught later on. After they had drawn the qu. ess. Lunae out of the jar, in the form of living quicksilver, they preserved it till they needed it, and the earth of the silver stayed blended with the tartar and the sal ammoniac in the jar.

Then they took common distilled water, poured it into the jar, and kept it in a hot balneum, shaken or stirred every day 3 or 4 times. Tartar and ✳ dissolved both into clear water, but the earth or *corpus* of ☽ settled at the bottom of the jar. Then they let it cool down, poured the water off into another jar, drew it off *per alembicum*, and found the tartar and the ✳ at the bottom, each by itself, as good as they were before; and they can again be used for other purposes as before, likewise the vinegar, while they are better than before.

Then they took the earth of the silver, dried it over a small fire, powdered it finely, and set it to calcinate

for 3 days in a reverberation furnace, fairly hot - only
red and no more, so that it should not turn into glass.
If one kept a strong heat, such as is used for forging
iron, it would turn into glass, for all things are by
nature ultimately glass, when the spirit has left them.
When it had thus reverberated for 3 days, they removed
it, put it into a stone jar, poured distilled vinegar on
it, mixed it well, and set it in the balneum for 8 days,
stirring every day 3 or 4 times and keeping the balneum
steadily boiling hot.

After the 8 days they let the balneum cool down,
poured the vinegar sideways off from the feces by turning
the jar to one side. After they had drawn the vinegar
off, they found at the bottom the clear, white silversalt. They put this salt into the reverberation furnace
for 24 hours, as mentioned before. Afterwards they put it
back into the jar, poured distilled vinegar on it and set
it in the balneum, proceeding in every way as previously.
When the vinegar was drawn off, they found the sal Lunae
at the bottom, glistening like ice and melting like wax.
They preserved it for the time they would need it.

Now they took all salt, and the 6th part of ☿ of ☽
and rubbed them on a stone till one no longer saw any

thing alive of ☿. Then they put it into a small glass
(No. 20 pg 34) sealed it, and set it in a dish with ashes

on a gentle fire, for 24 hours. Within this time the ☿
turned into a hard mass. Now they broke the glass and

powdered the mass finely. Then they took the other part

of ☿ ☽ae and mixed it with the mass, so that one could

see nothing more of it. They put it into a glass, to boil

in ashes, as before. They repeated this till all 6 parts

of ☿ were imbued with its own salt, and the corpus, or

salt, had drunk all its spirit. And this is what *Morienus*

says: Refresh the thirsty with living water, and he will

no more feel thirst in all eternity. Afterwards they

powdered the mass very finely, put it into a fixation-

glass, and poured rectified aqua vitae on it, one hand's

width above the matter. Then they sealed it and put it

into a dish with ashes to boil, first with little fire,

and in 12 days everything was dissolved into clear water.

Then one can see the spirits rise with the bodies, and

the aqua vitae as silver streaks; and when the streaks

begin to become coarse and thick, you must increase your

fire, and 24 days afterwards everything will be fixed

with the aqua vitae in the form of a glossy oil. When

that is cold, it will be somewhat thicker, like oil or

honey, clear transparent like crystal. It is a perfect

elixir. Put it into a glass *ampulla*, closed above, and

put it into the secret furnace of the philosophers, or *in*

tripode. That is the dry furnace of the philosophers.

Give it moderate heat, so that you can keep your hand in

it without burning for the duration of one *Ave Maria*.

Then the elixir will be congealed into the philosopher's

stone within 40 days, which will change all imperfect

metals and well washed ☿into true, better and nobler ☽

than comes out of the mines. *Raimundus Lullius* speaks of this work that it is a stone and a medicine, to which no foreign matters are added, but that one takes away from it what is too much. Praise God!

CHAPTER LI

Understand that you can make an elixir from ♃, and from ♄ alone to the red, for all old philosophers unanimously agree that ♃ is at bottom fine ☽, but yet raw and not boiled enough by the heat of the sun.

Likewise they say ♄ is fine gold in its innermost, but impure. If its impurity were taken away from it and its innermost turned outside, it would be perfect gold. You must not for this purpose open your purse and incur much expense in the Art, for God, who has made all things without a diminution of his essence, has prepared enough matter for us to fulfill all our desires, provided we know them and can prepare them. Let us, therefore, pray God that he may give us the intelligence to recognize the things which he has created for us, and that we may prepare it in such a way as to reach our goal, and to acquire the stone, for which we work, and to use the same for the honor of God and ours, and also for the happiness of the souls of our neighbors (actually: for the "salvation" or "heavenly bliss").

CHAPTER LII

THE TENTH WORK - *Aqua Philosophorum*

FOR THE WHITE

Some took 1 lb of ☿ of ☽, and also the salt of ☽ prepared in everything as has been taught in the ninth work; and they kept them for themselves. Further, they burnt an *aqua fort* of 3 parts, alum, arsenic, egg-chalk, calamine ana 1 part, saltpetre as much as all other parts weigh together. From this they made an aqua fort, pouring it each time anew on its feces, and again burning it 4 or 5 times. In this water they dissolved the ☿ of ☽, and drew the water off by distillation. They sublimated the ☿☽ae. They did this sublimation 3 times, each time with fresh water. Thus they drenched the ☿☽ae in a white spirit or tincture, and by dissolving and sublimating they made it subtle and increased it in its power and projection 100 times. Afterwards they took the prepared salt of ☽ and dissolved it in rectified aqua vitae; drawn off 5 times of *ana* alum, arsenic, egg—chalk and calamine. They again distilled the water off the *sal Lunae* in a lukewarm balneum. Upon the salt they again poured water till it was dissolved; and again drew the water through the balneum. They dissolved and congealed with fresh aqua vitae till the silversalt no longer congealed but stayed like oil, white and glistening like snow.

After this they took the ☿ which they had sublimated, rubbed it on a stone, and imbibed it with itself. That is what the philosopher *Danthin* says: Give it to drink its sweat, and it will become strong, so that one cannot overcome it in all eternity, for it will henceforth no more lose its power, that is, one will be unable to rob it of its sweat, because they are now simple (or: "one") like the insurmountable heaven, and it is no longer ☽ but qu. ess.

After they had imbibed all oil and salt into ☿, they took a glass with a long neck (No. 21 page 75). Into that they put the matter, sealed it above and put it for 8 days on warm ashes. Then the matter became hard as a stone. Now they broke the glass and powdered the matter finely; they put it into a fixation-glass and poured 1 hand's width of aqua vitae upon it. They sealed the glass and set it on a furnace, in a dish with ashes. First they gave a gentle fire, as if one wished to burn rosewater. Within 10 days they saw the matter with the aqua vitae rise, like silver streaklets, up and down through the fixation-glass, and when the streaklets began to become coarser and thicker, they increased their fire considerably, for your matter starts becoming coarse and fixed. And that is what Hermes says in the regimen of the fixation: It rises from the earth to heaven, and again falls down upon the earth, and the nethermost turns into the uppermost, the uppermost into the nethermost. And when the matter begins to thicken, you may be assured that the end is near, and the fewer veins you see in the glass, the more you must gradually increase your fire. They continued with this regulating of the fire till they saw no more little veins rise and descend. And they saw the matter at the bottom of the vessel like a bright, glistening oil, white as crystal. And it was a perfect elixir, penetrating the ☿ and all leprous metals into real silver, withstanding all tests, better than what comes out of the mines, (or: the ores). If then they wished to turn this elixir into the philosopher's stone, they put it into a glass *ampulla*, closed above and set it into the furnace of the philosophers or *in tripods*, that is the furnace of the philosophers or the dry furnace, in which they used to calcinate and congeal their spirits; and they kept it in such heat that they could hold their hand without burning between the jar with their ampulla and the furnace. They let it stand in this heat for 40 days, in which time the elixir had congealed into a hard crystalline pulverizable stone. And it grew 100 times in the first projection, and it is the philosopher's stone. Say thanks to God Almighty, be kind to the poor, and work sufficiently.

CHAPTER LIII

THE ELEVENTH WORK

Others took ♃ and ☽, and dissolved them in the aforementioned ▽. After that, they sublimated them, in every way as in the work just related. They also prepared the salt of ☽ just as before; thereafter they dissolved and imbibed the *sal lunae* with its own spirit, as in the previous work the salt with the ☿ of ☽. This is what *Morienus* says in the book of the Mineral Stone: Refresh the earth with the heavenly dew, and the earth will become fertile and produce flowers of a heavenly color. And after they had imbibed the sublimated and dissolved ☿ in its own salt, they prepared it in every way as I have taught in the above work; and thus they turned it into the philosopher's stone, into the same power and projection. In this way the old philosophers made their elixir in many ways, and prepared their stone and immediately brought it into very high power.

CHAPTER LIV
THE TWELFTH WORK

There were others who calcinated ☉ and ☽, and dissolved the chalk in balneo in distilled vinegar, till everything was dissolved into pure water. Then they drew the vinegar off by distillation till a small skin appeared. Now they removed the helm and put it into a glass pot, into a very cool cellar, for 5 or 6 days. During this time small stones sprouted, nothing else but crystals, from ☉ as well as from ☽. They removed these little stones and again distilled some water off them; and put them again into the cellar. They repeated this till nothing sprouted; then they put the crystals into a glass pot with a helm, set it in a cupel with ashes, distilling with a moderate fire till nothing dripped.

Thus the ☉ and ☽ remained so to speak as if it were oil; as soon as it cooled down, it was hard like a pulverizable stone, which they did in fact pulverize very finely. They dissolved it in distilled vinegar, as before, and drew off the vinegar till a little skin appeared, and set it in the cellar to crystallize as be-fore. They repeated this till no more feces remained, and thus they had purified the ☉ and ☽ from their uncleanliness.

Then they took these little stones thus purified, pounded them to a subtle powder, and put that into a fixation-glass. They poured a hand's width of rectified Aqua vitae on it, and proceeded with a moderate fire, as has been reported above, till it stood like a fixed oil and was a perfect elixir, which *transferred* ☿, ♃ and ♄ into ☉ or ☽, according to what had been the *ferment*.

If they wished to change this elixir into the philosopher's stone, they put it into a glass *ampulla*, set it *in tripode* for 40 days, and regulated the fire as has been taught above. Within the 40 days the elixir changed into the philosopher's stone. This is the Stone of which the philosopher *Calid* says that nothing foreign comes into it, only that one removes from it its impurity and urges it a bit so that it stays in ☿, that is in the Aqua vitae, which is wonderfully fixed with it in the fixation and is a means between (or: a medium, or intermediary) between the ☿ and the metals, and the philosopher's stone. For if the medium did not exist, the ☿ would in no way stay with the metals; but when now this medium is fixed with the philosopher's stone, the spirits remain constantly in the metals. In this Art many expenses are incurred and much labor is performed. Then, when they have worked for a long time and believe to have done their work well, after they have made it fixed, subtle and fusible and at last arrive at projection, they don't accomplish anything and are then as far as they were before. Then they think that the Art is impossible to practice, for they do not know the medium and do not understand the stone which God has given us for nothing. Neither do they know that by this medium all things must be joined, and because they have neither knowledge nor understanding of this medium, they can never reach *perfection*, even if they labored till Doomsday; but if they knew this medium, all their work would succeed. With this you cannot make any *projection* except on ☿ and ♃, and it is a Stone to the White; but if it has been prepared from ☉, throw it on ♄ and ☿, you will be taught the quantity of the projection by experience.

CHAPTER LV

THE THIRTEENTH WORK

Others make Aqua fort of Roman ♄ and ☽ to dissolve the ☉ ; or of alum de roche and saltpetre to dissolve ☽. They make such Aqua fort as is customary, and they dissolved therein ☉ and ☽, each by itself, because they wished to make both stones. After that they put each *a part* into a glass pot with a helm on it, and distilled 2 parts of the Aqua fort in balneo. Then they removed the helms and put the pots for 8 days in a cool, humid cellar. Thus, little stones formed in time at the bottom. Those they took out, put the helm on again and distilled each almost entirely in balneo. They set the rest once again to sprout, as above, and again crystals formed. They added those to the first, each to its kind. They cleaned the pots and put in it its stones, put the helm on, and set it for 8 days into a boiling balneum. During this time all these little stones congealed into a hard, pulverizable stone. Then they took the stones and pounded each by itself to a subtle powder; and they put each into a fixation-glass, and poured 1 hand's width of our Aqua vitae, prepared specially for each. They sealed the glasses hermetice, and boiled with such a regulation of the fire as has been taught in various places, till they stayed in the consistency of a fixed oil, red, transparent, clear like a ruby, to the Red; and transparent, clear, snowwhite, glistening like a crystal, to the White. And both were elixirs that tinged ☿ and ♄

to the Red, and ☿ and ♃ to the White. But if they
wished to transform both elixirs into the philosopher's
stone, they put each elixir into a glass ampulla, sealed
them hermetice and put them into the furnace of the
philosophers for 40 days and nights with such a regulation of the fire as has been taught previously in various
places. In the given time they turned into the
philosopher's stone, hard, transparent like a ruby to the
Red; and snowwhite, crystalline to the White.

My child, Thank God and work diligently, for what you
are here taught of these two stones, you can make from

♄ to the Red, and from ♃ to the White, although I
believe that it would be done more easily and perfectly

from ☉ and ☽. Nevertheless, all philosophers and old
folks concur that lead is leprous gold, and that it lacks
nothing but that its impurity be removed from it, and its
inside be turned outside and its outside, inside; then it

would be fine ☉. That is why it is known, for the
reasons mentioned, that the philosophical stone, white
and red, can be made from these metals without a foreign
addition, provided they be cleansed of their impurity and
joined to an Aqua vitae, which is the medium.

CHAPTER LVI

THE FOURTEENTH WORK

Take *antimonium*, sublimate it by itself. Then make an Aqua fort from Roman ⍺, 1 lb; saltpetre, 2 lbs; salt, 1 lb. Pour this ▽ 4 finger's width above the sublimated ☿, distill it off again in baineo till it is quite dry. Add to it half as much ✳, and sublimate them together with a strong sublimation fire. Put what rises separately into a glass, well closed so that no air can escape; what has not sublimated of the ☿, pound small, and pour water on it which has first been distilled of it; put it in the balneum to distill till it is dry.

To this add again half as much ✳ and sublimate it as before. Repeat this till all ☿ has been sublimated. This is called the Qu. Ess. of the antimonium. Now take this qu. ess. antimonii, put it into a glass, well closed with glue, sand and wax. Set it in a hot steam bath of water for 3 days and nights. It will dissolve as white as milk. After this, lute a helm on it and separate the four elements. First distill the bad water from it with a gentle fire and throw it away. Put another recipient on and let the balneum boil. Then a white water will go over it, which is the element of the water or *lac virginis*. Distill till it no longer drops; then distill it out of the ashes, and keep that also, for it belongs to the white water. Such tinges tin and copper into fine, good silver.

After this, pound what remained in the glass very finely and set it to dissolve in putrefaction, as has been taught above. Then an oil will appear, pale as water. That is the element air. Now give stronger fire in the ashes, and a red matter will appear, like thin blood. Increase your fire again till all matter has gone over. Then you have the element fire. The element earth, however, which stays in the glass, is good only for throwing sway.

Now take the first element of the water, *lac virginis*; put it into a glass and close it hermetice. Set it into the athanor which one has in the warm stove, and give a very gentle fire. Then all colors will appear, for after the blackness, a beautiful whiteness will come. Then it has enough. All this will take place in one month and it will be a tincture for silver, which mixes with

♃ , just as all silver with gold.

Now put 1 lb of ☿ into a crucible; heat it till it starts smoking. Pour upon it one "Lot" (half an ounce) of the said elixir, and give strong sublimation fire. Then it
will stand as if the crucible would break. But if it

becomes brittle, add more red ☿, and you will have very

fine ☽, steadfast in all tests. The element of the fire, however, which you have kept apart, close it into a glass as previously the white water, till finally a red color appears. This tinges, in everything as in the white, the ☿ into good and fine ☉ which stands all tests, better than what comes from the mines.

CHAPTER LVII

THE FIFTEENTH WORK

This is an *elixir* from ♂ and ♀; how the golden shirt is to be pulled off the black king, which is ♂, and is to tinge ☽ into ☉.

Take the black king and pulverize it, and mix him with pulverized tartar; put it into an alembic, and first distill of it the tartar. Preserve this water well stoppered. As to the powder that stayed below in the vessel, put it into a calcination—vessel and calcinate it in the wind—furnace, or calcinate it for 3 days and 3 nights. When it is well calcinated and has cooled down, pour good distilled vinegar upon it and let it stand till the vinegar is colored yellow-red. Then pour it off into another alembic, and again pour other vinegar upon it. Let it stand again till it is colored, and then pour it off carefully to the other red vinegar. Again pour fresh vinegar upon it, and do this till all redness has been extracted from the chalk.

Now take all the tinged vinegar and draw it off per balneum Mar. You will find a red powder at the bottom of the alernbic; that is the qu. ess. of the ♂̸; dissolve this powder in the cellar upon a glass tablet to the consistency of oil.

Pour to this oil its own water which you had first drawn off, and put it into a retort with an alembic; or lute another retort over it, the joints (seams) being well closed. Set it in the furnace on warm ashes; let it rise and descend, and you have a fixed and incombustible oil which fixes and tinges ☽ into gold. Take thin plates of fine silver, burn them and let them go out, in this oil, and they will turn into fine ☉ This oil cures all man's sicknesses and all wounds, including leprosy, fever, fistulas, and cancer, etc.

CHAPTER LVIII
THE SIXTEENTH WORK

This is another elixir from ♂̸ for all sicknesses.

Take *antimonium*, pound it well with distilled vinegar. Put it in a warm spot to dry. Again pound it with vinegar and let it dry. Then take powdered salt as fusible as butter. Sublimate the antimonium through it (or: with it, or: by it) for 14 hours, first with a gentle fire, then gradually increasing it. When it has turned cold, break it open, and you will find the antimonium as white as snow, sublimated. Now pour warm water upon the feces that stayed at the bottom, filter it off as one would make a lye. Repeat this till all saltiness is gone from it. Dry what is left over and keep it. Then take same and pour distilled wine vinegar on it, so that it stands 3 finger's width above it. Stir it well and put it with the glass on warm ashes for 3 hours, and let the matter settle at the bottom. Then pour the vinegar carefully off; pour other vinegar upon it, stir it well, and let it stand in hot ashes as before. Do this 4 or 5 times.

Keep the drawn off vinegar in a glass, for in it is the qu. ess. of ☿ ij. Now take this vinegar and distill it off completely by Baln. Maria, and the qu. ess. stays below in the destillatorium. After this, take what has been sublimated and do likewise with it; draw its qu. ess. off. Then take it and let it putrefy in horse dung or in balneo till the oil is dissolved. Then it is sweet as honey. This oil has inexpressibly great virtues for chasing away all contagious diseases and to keep man healthy; it heals all wounds, outside and inside, and leprosy, etc. *Glossa*: I mean to say that one has to bring the qu. ess. and the sublimated together in an oil and use it, as has been taught above.

CHAPTER LIX

THE SEVENTEENTH WORK

HOW TO MAKE AN ELIXIR FROM 2007 AND 1111

Others took one part 2007 and melted it. When it was melted, they threw into it 3 parts of 1111, stirred them together and let it cool down. Then it is brittle and soft. Pound it and wash it very well with salt and vinegar till no further blackness gets off it. Then dry it at the sun or a small fire. After that put it into a glass retort and, in addition, twice as much as its own of the Aqua fort described hereafter. Distill it up and down till it stays at the bottom like an oil, quite red, and it can no longer be congealed. Then you have the Aqua vitae and oleum philosophorum incombustibile.

Now take of this oil ℥ ij, or as much as you wish; add to it 1 Lot (half an ounce) of thinly beaten, fine goldleaves. Put them together into a glass retort or *phiole*, well closed, and let it stand for 7 days and nights on a moderate fire.

Then everything together turns into an oil. Now add again 1 Lot of fine gold, and let it again stand for 7 days and nights in warmth, as before. Then it becomes as thick as molten pitch. Now add 1 Lot of gold as before, and thus there are 3 Lots of gold in 4 Lots of oil. Let it again stand 7 days and nights as before, and you have a meltable medicine, like wax, to the utmost Red. It turns stiff in cold air. You must pulverize this, and throw 1 part on 500 parts of fine silver; it will all turn into fine ☉

The aforementioned water is made in the following way: Take Roman ♑ 1 lb; alumen roche and common salt ana, 1 lb. Powder them small and mix it with 1 lb of distilled wine vinegar. Of that burn an Aqua fort; into that put 1 ℥ ✳.

CHAPTER LX
THE EIGHTEENTH WORK
ABOUT TWO WATERS

Now, my child, I will teach you to make two waters with which one does wonderful works of Art, for without these two waters, no one will draw the Stone from ☿ alone. *Arnoldus de Villanova*, *Raimundus* and *Albertus Magnus* have considerably improved this water, since they found each time more truth in it. Arnoldus found that one should add to it *crocus* and *lap. haematitem*, an equal measure of each. Raimundus found that one should add *antimonium minerale* and *vermillion*. Albertus Magnus found that one should put in it *aes usturn* and "Spanischgrun" (basic copper acetate or carbonate). All old philosophers were sceptical in this regard and made their Aqua fort from Roman vitriol and saltpetre, or from alum and saltpetre.

106

That is also why it took so long till they accomplished anything that was perfect. That is why I will now teach you to make the first water, of which there was mention made before in the Mineral-Stone. It makes the stone volatile. Make it as follows:

Take Roman vitriol, 3 parts; antimonium minerale, 2 parts; lap. haematite, crocus Martis, aes ustus, Spanischgrun and cinnabar, 1 part of each; saltpetre, 10 parts. Dry them to a powder and burn an Aqua fort, first with a gentle fire for 24 hours; subsequently with a stronger fire. When it has cooled down, remove the caput mortuum from the jar, powder it small, and put it back again into the distillation-vessel or the jar. Pour Aqua fort upon it. Immediately put a helm on with a recipient, closely luted, and distill as before. Do this 3 times, each time pouring the water on its powdered feces; and drawing it off again. This is then called *aqua philosophorum* for the Red, which you will often be ordered to use in the Art.

CHAPTER LXI

Now, how also to make the other water which gets fixed with the work, of which there has been mention made in the aforementioned Mineral Work, when I said that it is made with danger and with understanding of the alchemist. Proceed then as follows: Take Roman vitriol, that is very pure and transparent, and vermillion, of each 1 part; crocus Martis, lapis haematites, aes ustum,

Spanischgrun, of each ½ part. Dry and congeal them to a dust, till it does not retain any spirits. If half a part of antimonium minerale were also required, one should extract its spirit with vinegar in a lukewarm balneum, so that the spirit of antimonium is congealed before it is added to the work. But I will later teach how the spirit is to be drawn out.

Now put the aforementioned matter, that has been dried to dusting, into a recipient; pour on it as much rectified Aqua vitae as was drawn over still 12 or 14 times after the rectification. Quickly stopper the *receptacul*, so that no spirits can fly out. Put it for 7 or 8 days in a warm balneum; then take it out, add a helm with a recipient, and provide yourself with such lute that you do not lose your spirits during the distillation; otherwise your work would be spoilt. Distill with little fire for 2 days and nights. After that, increase your fire for another 2 days. Following this, keep your glass with the matter burning for 3 days. Then let it cool down, remove the helm, and break the glass in order to remove the caput mortum. Powder that finely and put it into the recipient in which your water is. Close it tightly and set it for 8 days in the balneum, in such a way that you can just suffer your hand in the balneum. Now take it out and put it on the fire, gently heating the first day, more strongly the second; the third day, so that it glows (burns), and let it stand thus for 24 hours and afterwards cool down.

Now put again fresh matter, like the first, in the water and proceed with it as before. Do this 3 times, using fresh matter each time. You may also prepare this water in earthenware jugs, which are made at Siburg near Cologne. Add a helm on top that has a hole at the crown to put the matter inside; for when the helm is very firmly luted on it and the spout at the end is well blocked with wax, you do not lose any spirits. For if you were to lose the spirits, everything would be spoilt, since the tinctures are in the spirits and not in the bodies. In such a way it must be prepared with great care and danger, and with the alchemist's intelligence, so that one does not lose the spirits, otherwise your work would be spoilt.

CHAPTER LXII

My child, I am telling you with right love that never has there been found a greater secret in nature than these waters, since with them one can accomplish all the kinds of works that can be done in *alchymy*. Because of this second water there have been so many errors in the Art *Alchymia*; yes, all our forefathers have so far not known of this second water. That is why it took them so much time till they accomplished anything perfect, and they worked also with great labor, for long times, and special care. Here, however, my child, it is explained to you with clear intelligence and labor. Therefore, thank God the Almighty that the first of these two waters is called *aqua philosophorum*; we call the second our burning, blessed water. That is why, when I say aqua philosophorum, I refer to the first water which contains the saltpetre that makes the Stone volatile. And when I speak of our burning water, I mean the second, which gets fixed with the work (or: coagulates). Therefore, my child, be very careful not to mistake one for the other (or: not to use one instead of the other).

CHAPTER LXIII

THE NINETEENTH WORK CONCERNING ☿

Now take quicksilver, 4 or 8 lbs; or as much as there was quicksilver in the first two sublimations; but dry your ⊕ without losing the spirits, also the common salt, till they are like dust. Mix them and grind them on a stone with strong, distilled wine vinegar, till the quicksilver is no longer noticed - as subtly as if one were to paint with it with a brush. Now put it into a stone pot made for that purpose; add a helm, luted firmly and tightly; set it on a furnace in a cupel with ashes or sand; lute a recipient firmly to the spout; draw the wateriness up with a gentle fire till you see that the helm gets dry and nothing drips from the tubes of the helm. Now increase your fire, so that it can nicely sublimate; when the fire begins to throw flames, add more coal, so as to keep it for 24 hours in even heat. After this, increase your fire considerably till the bottom of the pot begins to glow nicely; let it stand for 6 hours in a gentle glow (heat). After this, make it burn more strongly, in order to drive up the coarse spirits of the vitriol together with the quicksilver; since the volatile part of the vitriol is embraced by the ☿, because it is its like and they are of the same nature; that is why they clasp each other. Therefore, *Aristoteles* says in the Book of the Stone, in the 8th: The spirits that are quite fixed are of no use as long as they are earth, and they do not notice that the philosopher's stone, which God has given us for nothing, that is, vitriol, is fixed in one part; otherwise it would have no nature with the ☿ and the metals. For in its unfixed nature it receives the ☿ in its living being alive; and when it has received the same, they become together one corpus and congealed quicksilver, which was previously alive. And thus the stone which God has given us is united with the quicksilver, and they become one *corpus*, which is volatile, and likewise with the fixed part.

Afterwards, when one has made it to rise and again fall, one congeals (gets fixed) with the other, keeping their natural moisture. Thus it does not lose its natural *ingress*, as *Geber* says. Whoever sublimates otherwise than through the stone that God has given us for nothing, will never attain to the righteous and perfect Art.

CHAPTER LXIV

My child, I am telling you this so that you should know what you are doing, and understand what is *sublimation*; so that you do not imitate the blind who wants to lead another blind. You must know that you must always sublimate at least once or twice through the feces, in order to rise the more fixed part together with them, for *Morienus* says: If one part is volatile and the best part is fixed, it retains the best, which is volatile.

My child must know that in the first sublimation one takes 2 parts of Roman vitriol to 1 part; but in the second sublimation, or after the first, one takes as much of one as of the other, and one sublimates twice through the feces. That makes 3 sublimations. My child must know that in the first sublimation one takes twice as much

Roman vitriol as ☿, because the ☿ is alive; for if you

did not take more Roman vitriol than ☿, the ☿ would not congeal. Neither could you grind it dead on a stone, but it would always live and not mix with the vinegar. Therefore, if you grind it, grind it such that nothing

living is seen of the ☿ on the stone and it is totally killed; otherwise your sublimation would be good for nothing, and your spirits of the Roman vitriol would not rise with the quicksilver, nor would they grasp each other thoroughly, but each would sublimate and rise by itself. Sublimate it always once or twice through its feces, till all the quicksilver is dead and hard, so that you can powder it before you sublimate it again through fresh vitriol. After that, do not take more of one than of the other.

If the quicksilver were not yet dead, you would now not be able to kill it, because there is no more of the one than of the other. Therefore, take good care to incorporate it well on the stone the first time, and to grind it diligently. If you have to mix it again with its feces, do it with good distilled vinegar, as finely as if one were to paint with a brush with it. In this way you must always proceed when you wish to sublimate ☿ through Roman vitriol. If, however, you wish to sublimate it without any addition, pulverize it dry and always set it dry to sublimate. Know also that you must well sublimate it 5 or 6 times before you dissolve it, that is, the first (sublimation) 3 times through double the weight; and then 3 or 4 times through equal weight. My child, I have now taught you sublimation. Take care that you memorize it well, for great errors occur during sublimation; when it is done poorly, the work is spoilt.

There are many who have the audacity to sublimate and do not know what they are doing; they believe that they have well done their job. Then, when they find nothing fruitful at the end, they consider the Art impossible. For them the Art is indeed impossible, because they understand neither themselves, nor the Art, nor the work at hand. They wish to sublimate and do not know why they sublimate quicksilver with vitriol. They know nothing else but that one has to sublimate in order to rubify; but they ignore that it must be joined to the stone which God has given us for nothing. And if it were not sublimated with that, one could never unite it with the metals, for without this means, it would not stay in the metals; for the stone that God has given us for nothing is a medium between the metals and quicksilver, as said before. Because they neither know nor understand this medium, also because of other reasons concerning the Art, they can never reach *perfection*. Therefore, open your eyes and see; and your ears, and hear; and open your intelligence, for here you are shown everything you need - with this, enough of sublimating.

CHAPTER LXV

Now we will deal with *solution*. After the quicksilver has been sublimated, dissolve it first in aqua philosophorum; and thus dissolved, put it into the sublimation-pot; put a helm on, and a recipient, everything well luted; and distill it first with a gentle fire, till all wateriness has gone over and you see the quicksilver sublimate in the helm. Now increase your fire a little, so that it continues nicely to sublimate for another 6 hours. At the end of the 6th hour, make the bottom glowing. Let the pot stand in the heat for 4 times 6 hours, which together is 24 hours. When your water has been drawn off, let it cool down, remove the helm and the (??? - blank in original). The water that is in the recipient is no longer any good, or one may keep it to calcinate certain bodies, otherwise it is no good. For the *argentum vivum* has carried with it all power and all tincture, because they are equal. From a horse comes a horse, from a dog, a dog, etc. All tinctures in the water are like, and they have all received their original

nature and first origin in the mines from ☿, and he is

the father of all. Thus the ☿ carries all tinctures to
his nature about with him, and he marries them only in
the solution, for all natures have turned into water and,

as ☿ at the same time becomes water, they all become one
and joined in the solution. Thus there is a marriage

between the tincture and ☿, which marriage is later,
after the fixation, to be considered the matrimonial
work. That is why *Aristoteles* says: I do not believe that
one can transpose one thing out of its nature into
another, unless one first bring same into its first
nature; then it is possible to change it into another
nature. All things were originally water. Therefore, if
someone wants some *perfection*, or wishes to make an
elixir or the lapis philosophorum, he must needs turn
everything together into water before.

CHAPTER LXVI

Now take the quicksilver and dissolve it again in
fresh *aqua philosophorum*. Again draw the water off
through the helm, with a gentle fire, as before. Do this
at least 5 or 6 times, or at the very least 4 times, each
time resublimating, and take each time fresh aqua
philosophorum; for it must have so much tincture that
within it is as beautiful as fine gold. My child must

know that ☿ becomes so subtle and strong in every solution and sublimation, that it increases each time in projection a hundredfold when it has been brought to the elixir or the philosopher's stone. That is why, my child, do not tire of dissolving and sublimating, for the time and labor is paid and rewarded a thousand times.

CHAPTER LXVII

When everything has been dissolved and sublimated, sublimate ☿ 3, 4 or 5 times, till it leaves no more feces, for that is what *Morienus* says: Unless you take its blackness away and make it clear like a crystal, you have not accomplished anything in the magistery; therefore, sublimate it till the bottom of the pot stays as beautiful as it was when you put it into it, and it is clear like a crystal. Then it is prepared up to the calcination. Therefore a philosopher says: Christ is taken from the cross and put into the grave.

CHAPTER LXVIII

Now they took the quicksilver, prepared in this way, and closed it in a glass ampulla or in a philosophical egg, which they filled entirely, so that the spirits should not rise. They ground it very finely on a marble or glass plate before putting it in (the egg). When it was inside, they stoppered it tightly, after they had thus packed it as tightly as they could, so that the

116

spirits should have no air; for if they had air and the heat were too great, the spirits would rise in the glass. That is why they packed it as firm as they could, filled to the brim, so that nothing stayed empty. Then they sealed it hermetice, set it for 6 or 7 weeks in tripode and heated so strongly with coal that they could hardly hold their hand between the wall of the furnace and the dry stove in which ☿ stood calcinating.

As soon as they had checked the heat by looking through the square hole of the big furnace, they closed it firmly with a stone that fitted into it, so as to retain the heat in it. During the day and the night they inserted their hand 4 or 5 times through the square hole, so as to feel if the heat was suitable. It must stand thus at least for 6 weeks, in order to well digest, or calcinate, or open up, the matter; for to digest is as much as to digest (Note: "digerieren" is a latinized expression which means the same as "verdauen".), like the food in the pit of the stomach. The stomach is the vessel in which the food is boiling; but the liver, which is spread under the stomach-pit, must open the food with its heat and dryness and separate the elements, each for its type of food. In the same way this *materi* must also be digested and opened with the help of a moderate fire, so as afterwards to separate the elements from it, so that each imperfect metal should take its nourishment, each what it requires to reach health and perfection. That is why, my child, keep your fire moderate or your work is lost. For if you were to make it too hot, it would dry itself up. Afterwards you could not get any moisture or water and could not *albify* (make white), and thus your work would be spoilt. For if you cannot have any water, with what will you have fire and air, for such must be done by means of the water, just as the preparation of the earth. Consequently, water is the axe and the hammer with which the workman performs his work, and afterwards shows it up.

Such it is also with this water. With it you must separate your elements and prepare your earth in order to perfect your work. After that, however, the water does not remain with the work, as will be taught later. Therefore, my child, do not make your fire too hot, so that you do not dry up your matter. Neither give it too gentle a fire, or else the matter would become too coarse, raw, and remain closed, and would not open up to get the water thereof; neither could it be separated.

Note: When the food gets into the stomach and the liver is too cold, it is not digested and the food remains closed and unconsumed, which people throw out and spit out. Should it happen that by chance or negligence you have made your fire too hot or too cold, or if you left the matter stand too short or too long a time *in tripode*, and then you would get to the stage of distillation but the matter would not rise, being too closed or too dry; or if you had taken it out too early and it would thus not be subtle enough, or not suf-ficiently opened, and would not dissolve in the *aqua philosophorum*, as you had done before — then draw the

118

water off and sublimate it as before; put it again in a glass ampulla or an egg of the philosopher's, and then in tripode, and give fire as before, and take better care.

My child, with this I have taught enough of *calcination*, *digestion* or opening. Do not begrudge the time you spend on the work and understand each thing correctly before you begin to work with it, etc.

CHAPTER LXIX

When the 40 days and nights are over, remove ☿ from the tripode and break the glass, because the matter is hard and baked together. Take it out and grate it on a stone or glass plate, very finely. Put the powder into a glass pot, lute a helm on it with lute that can stand water, and set it in the balneum of this shape (No. 22). You must have a long kettle, 2 ells long and a half ell wide. At the side there must be a pipe for pouring in boiling hot water, when occasionally something is boiled away. Into this big kettle one has to hang a small kettle, half an ell deep, which lies with the rim on the long kettle and fits that well. Below, this small kettle must be full of small holes, through which the steam of the balneum can go round the retort (or: alembic) which stands in the small kettle. Set this long kettle into a furnace. It must be level with the kettle above at the rim and all around it 1 hand's width of room between the furnace and the kettle. This furnace should have two holes for ashes and above them, a grate (or: grid) through which the ashes fall. One and a half foot above

119

the grate there should be a hole for heating. Put a strong iron cross in it for the big kettle to rest on. It must be attached above with the uppermost layer. The stones of the uppermost layer of the furnace must be hewn, so that they fit all around the kettle and keep the heat within. The furnace has to be lined with strong glue, 1 hand's thick, so as to retain much heat - then it is ready.

Now hang the small kettle with the holes in the long kettle and put on the floor of same; small, subtle (or: fine) hay, or cowhair, 1 hand thick. Into that set the glass pot with the matter and the helm on it. All around the glass stuff fine hay, or cowhair. After this, make a round leaden lid (or: cover), which fits all around the small kettle and lies close by the glass pot. The reason is that in the center of the lead a round hole must be cut into which the glass alembic fits, so that no steam can escape anywhere except through the pipe of the long kettle into which the hot water is poured. When it is ready, put the helm on, lute the recipient to the spout, fill the kettle through the pipe with clean water, put fire under the furnace, and let the balneum boil. Keep it thus for 15 days and nights, always having hot water at hand to fill the balneum anew. Provided your matter has been well opened during calcination, and provided you have kept your fire at an even temperature, you will not distill beyond 9 or 10, or at most 12 days, since you will obtain enough water. Otherwise you will not get it for a long time, that is to say, hardly in one month or 6 weeks, or perhaps never; so that you are obliged to remove the matter again, dissolve it in aqua philosophorum, draw off the water, sublimate and grate it, and again put it back into the glass ampulla or the philosophical egg, and set it back in tripode, keeping the temperature of the fire just as above.

But it cannot be said for certain how long it takes to draw the water from the balneum over, nor how much water you must distill with which you have to dissolve all your substance or matter, since there is no specific measure or weight for it. Nevertheless, it is easy to have enough water for it, for when ☿ dissolves (and again congeals), it can well dissolve in common water; and once it is congealed, it is likewise easy for it to congeal (again). This is said by Hermes in the Book of the Stone: Thus some have drawn over 2, 2½, 3, 4, 5 parts of water, depending on whether their work, which they had begun, was small or big.

CHAPTER LXX

If it should happen that your matter does not dissolve, I will show you a way to dissolve it. Put the helm back on, again distill the water off, grind your matter with a little distilled water on a marble slate till it is like pap. Put it back into the glass, pour all the distilled water upon it, and put it into the balneum for 5 or 6 days. In time it will dissolve, and when you have again drawn off the water, remove the matter from the pot, grind it on a marble slate or on a glass slate,

and then put it in a glass (No. 23 page 75) with a round
belly and a long neck, luted tight above and turned in,
like a small pin (or: peg or plug), so , when the bottom
rises, it can again start dripping. On each side there
must be 2 arms rising from the belly to the neck, so that
the watery smoke that rises into the neck can escape,
Somewhere, (maybe in the neck) there is a glass tube,
like a funnel, through which the pounded matter has to be
put into the glass, well shaking it, so that it may fall
upon the bottom. It is also through such a tube that one
has to pour the water upon the matter. Afterwards, seal
the tube hermetice, set the vessel into the balneum,
keeping it thus till everything is dissolved into pure
water without feces at the bottom. This is the best kind
of a dissolving glass, because what steam rises from the
belly through the arms into the neck, drips at the point
which goes down from the neck,, and this helps quite a
bit with the solution on account of the steady dripping
day and night. I have not found a better kind to dissolve
quickly and well.

CHAPTER LXXI

When everything is well dissolved together, let the
balneum cool down, pour it into a distillation pot
through the tube through which you had poured in and
which was sealed. Put a helm on, put it on the furnace in
a dish with strained ashes, add a recipient, distill all
water and milk down with a moderate fire, and keep that
well stopped till you need it. Take your matter out,
pound it very finely on a marble-stone or glass tablet.

Thus pounded, put it into a glass ampulla or egg of the philosophers; fill it completely so that the spirits do not rise, and seal it hermetice. Now put it in tripode to calcinate, digest, or open, as before, for it must again be digested and then again dissolved. With the help of the water (or: by means of the water) the element air is to be drawn out. Be careful, therefore, to give the same heat as before, for 30 days and nights. Then let the furnace cool down, remove the egg with the matter, and break the glass.

Take the matter and put it again into the dissolving glass through the tube, after it has been pounded finely. Again, pour part of the element water upon it, as much as you consider necessary for dissolving the matter, which now dissolves more easily than the first time. Seal the tube hermetice and put it in the balneum into the small kettle with the holes. Stuff hay or cowhair around it, and let it dissolve, as before. After this, let the balneum cool down, open the tube, and pour the dissolved water into the distillation-cask; put the helm on, and put it into a dish with strained ashes on the furnace, together with a recipient luted thereon; or first draw the water in the balneum off before you set it in the ashes. This would be the best and surest means, as in the balneum nothing rises but the element water, while in the ashes occasionally something of the element air also goes over.

If one has made the fire somewhat too hot, one has to separate it again in the balneum; that is why it is safest first to draw off the water in the balneum and then remove it and put it in a dish with ashes on the furnace, and distill the air off with a good fire. The air goes over in the form of a golden oil, of a beautiful yellow color. Do this till nothing drips any longer. Pour this air or yellow oil into a glass ampulla and preserve it well stoppered till you need it. Now take your matter out of the distillation glass and again pound it intangibly. Put it into a philosophical egg, completely filled, so that the spirits do not rise from the fire; seal it hermetice, set it in tripode, heat as before, for 30 days. Then let it cool down, take it out and pound it intangibly, and put it into the dissolving glass to dissolve. Put it back into the balneum as before, till it is all water without a sediment.

Let it cool down, take it out, pour it into the distillation pot, put the helm on and a recipient with it, distill the water off in balneo, pour that to the first element of the water, or virgin's milk, set your glass on the furnace in the ashes, heating at first gently, gradually more strongly, till at last the matter is glowing in the pot. Now distill the element fire blood-red. It tinges glass and everything it touches, and it makes it pliable, which is a wonderful thing. When it no longer drips, heat for another 4 hours with great violence of the fire. Then let it cool down, remove your recipient, lute it closely, and preserve it carefully till you need it.

CHAPTER LXXII

Thus you have the 4 elements, each prepared specially, that is, virgin's milk or water, air, fire, and earth, separated from ☿. Remove the earth from the glass pot, pound it small on a stone, and put it into a vessel made as a covered dish of Siburgian earth (or: clay); or take two flat bowls that fit tightly upon each other. Lute them firmly with strong lute that can stand the fire, and place them in an arched furnace or reverberatory to calcinate there for 8 days in even heat. Then take it out and pound it intangib]y; put it into a dissolving-glass, pour on it of your element of the water, or virgin's milk — a large amount; yes, even if you poured all your water on it, it would be all the better. Close it tightly and let it stand in balneo till everything is dissolved into clear water without feces. Then let it cool down, take it out, pour it into the distillation pot, and that in the balneum together with a helm on it and a recipient attached to it. Distill the water off and let it cool down. Remove the recipient with the water or the virgin's milk, close it tight and keep it for times of need.

Remove the helm from the pot, and you will find the fixed ☿ at the bottom, clear as a crystal, which salt or earth is the nourisher and the foundation of the Stone, and, as some say, thus one does not require any ferment of gold or silver, because the earth or salt is a corpus from which the Spirits have been withdrawn and separated. It is said to be the origin and *sperma* of all metals, that in it is the true ferment, and that they do not require any other ferment. They take one half of the earth or salt, and the other half of the element of air, and all the water; they mix them well and put (the mixture) into the fixation glass, seal it hermetice, and set it on warm ashes. There they let it rise and go down again till everything is fixed and remains in a fixed oil. This they put into a glass ampulla, or a philosophical egg, seal it above, and put it in tripode for 60 days and nights at an even, moderate heat. During this time it is congealed into a crystalline, pulverizable and fusible stone.

Then they take the other half of the air and the earth, mix them with all the fire, and put it (the mixture) into a fixation glass; they seal it, put it on warm ashes, let it rise and fall till it is fixed and turns into a fixed oil. Then they take it out and put it into a glass ampulla or philosophical egg, seal it, and set it for 60 days and nights in tripode in moderate heat. During this time it congeals into a crystalline stone, red as blood, pulverizable and fusible like wax. Then they believed that the Stone was perfect to the Red and the White; but when they came to make projection, they found nothing and were cheated with their false *opinion*, believing that they did not require any ferment except salt and *terrae Mercurii*; because it is the beginning and origin of all metals, and they relied solely on it.

It is true that ☿ is the sperma and the beginning of all metals, and that in him everything is hidden; but he is no metal and has never been one. Example: The *sperma* of man is man's origin, and man's nature lies hidden in it; but it is raw and unboiled and has never yet been a human being or a corpus; how could a human being arise out of it; in it there is neither soul nor life, except through nature with the help of the mother and wet nurse, which must be done by length of time and moderate heat.

Thus it is also with ☿. It must be congealed in its mother, the *minera* (the ores), by means of the air of Sulphur, and boiled in moderate heat, over a period of time, and it will thus become a metal and a corpus. Consequently, according to the above—cited reasons, these people are cheated.

CHAPTER LXXIII

Now they had, or they kept, their ferment in reserve, both to the White and the Red. They put them into a fixation glass, each powdered fine by itself. To the Red they poured the blessed burning water; and to the White, the blessed water. They sealed both glasses, put them on both sides on top of a furnace in a big dish with ashes; they dissolved the ferment with the powder of the Stones in moderate heat, till our burning blessed water was congealed into a fixed oil, which was a perfect elixir. Then they removed same from the fixation glass, put each into an ampulla or philosophical egg, sealed it hermetice, and set it in tripode for 40 days and nights in moderate heat. During this time they congealed into the Philosopher's Stone to the Red and the White, both having an infinite capability. This stone you can at any time *multiply* with 7 parts of multiplied quicksilver and 1 part of this Stone. Put them together into a fixation glass; pour on them of our burning blessed water, and when this has been congealed together with our blessed water, put it into a philosophical egg, hermetice sealed, set in tripode for 40 days and nights. Then you have multiplied the Stone both to the White and the Red, in like power.

CHAPTER LXXIV
THE PREPARATION OF THE FERMENT

Take gold or silver, whichever you wish, for all is one manner and one work. Calcinate them with a subtle chalk, or laminate and dissolve them in our red or white burning water; which, white burning water I shall teach later on how to make. Draw it off again *in balneo*. Do this 3 times, the last time draw it off, and your ferment is ready.

CHAPTER LXXV

My child, when you have separated your elements, as before, from ☿, and the salt or earth has been prepared, take all the earth, air and fire, mix them together and keep your virgin's milk thereof. Weigh how much these three weigh together, and add to all 7 parts, 1 part of ferment; that is 7 *mixture* to 1 ferment. Mix them and put them into a fixation glass. Pour on it as much of our burning blessed water as everything together weighs; seal the fixation glass hermetice, put it on the furnace in a dish with strained ashes. Everything together will dissolve into water, and you will see them rise and fall together, as a philosopher speaks: The lower must be as the upper, and the upper as the lower, or you have not accomplished the magistry. Let it thus stand day and night till it remains at the bottom like a fixed oil. But before it becomes fixed, you will see all colors, yes, more than you could imagine. Then there occurs a marriage or wedding, that is, a union among the elements, and the ferment and the burning water, and the things that are in the burning water. When then the colors manifest, each shows its virtue, and the fixation glass suffers great discomfort, so that it stands there and trembles, because all opposing spirits are thus fighting each other, so that the fixation glass may well sometimes burst; and all who are in the room may die from the air; that is how poisonous the air is when all colors fight with each other. This is the reason why it is highly necessary that the fixation glass be rather thick.

Therefore *Aristoteles* says: I heard and saw my children quarrel, and it tore heaven, and I ran out of the world. When now they have quarreled enough they finally reach one color, and then, the matter starts to congeal, and the marriage is consummated. Therefore, keep it day and night in moderate heat till it is fixed, and take great care not to make the fire too hot. When then it stays at the bottom in a fixed oil, let it cool down, and.you have an oil that is thicker than honey. It is a perfect *Elixir*. Now remove it from the fixation glass, put it into an *ampulla* or egg, seal it, and set it *in tripode* for 40 days and nights, in even, moderate heat. Then the Philosopher's Stone will coagulate, which can be pulverized and which is fusible like wax, transmuting all impure metals and quicksilver into real gold, better than that which comes from the mines. Its *projection* and power are infinite.

My child, Thank God and be kind to the poor. The *projection* will be self-evident when you do it. Later on we will tell more about projection. Here the work of the quicksilver is finished. You will also learn many different kinds of manual operation in connection with this work. (Fixation glass, or fixing glass.)

CHAPTER LXXVI
THE TWENTIETH WORK

My child, I now wish to teach you how to make the burning water to the White. Take alum *de roche*, dry it over a gentle fire till it dusts, without losing its spirits; also galmeystone (or: calamine), egg chalk (egg calyx) sublimated to the White, as will be taught later, *ana*; pound them subtly (or: finely), put them into the distillation pot, pour *aqua vitae* rectified of its *phlegma* on them, put a helm on, and distill a water as I have taught about the burning water to the Red, which

130

congeals during the work. Pour it again on its feces, and distill it off. Then take fresh matter, pour your water on it, and do in everything as has been taught for the Red; preserve this water for when you need it.

My child, I will now teach you what you should do with the element of water or virgin's milk, which you drew off before during the work of ☿ and which I told you to keep and which I did not wish to have for the work to the Red. There are some who put the air together with the earth, the virgin's milk and the ferment, but all this air is no good except in the White work, because it increases its tincture to the White. I will, however, teach you here another way of utilizing the water or virgin's milk.

Take fine cupel—silver, dissolve it in *aqua fort*, made from alum and saltpetre, *precipitate* it and wash the chalk of the aqua fort with common water. Dry it and put it into a jar; pour distilled wine vinegar upon it, ✳ and calcinated tartar, and proceed just as I taught you before. Draw it off again, and distill the ☿ of the silver, 1 lb or ij as above. Prepare the salt or the earth of the silver everything as I have taught above.

Now take the ☿ or the Qu. Ess. of the silver, sublimate it through 4 parts of alum and 2 parts of common salt. Dry the alum so that it keeps its spirits; mix them together and pound them on a stone together with the ☿

of ☽ and distilled vinegar till one no longer sees the ☿

☽ ae. Then put it into our sublimation, cask, and
sublimate it as has been taught above. After this,
sublimate it again through its feces; then sublimate it
again 4 times, always with fresh matter. Afterwards,
dissolve it in our white burning blessed water, and again
draw off the water together with the fire; and again

sublimate the ☿. Repeat this 4 times, each time with
fresh white-burning water, then the of silver of the *Qu.
Ess. Lunae* is prepared.

Now take the prepared salt of the silver or the

earth, which I told you to keep. Pound it with the ☿ of

☽, each by itself on a stone till it is intangible.
Further, take the virgin's milk, which I told you to
keep, and imbibe it by grinding together on a stone (or:

mixing) into the sublimated ☿ and salt of the silver; or

put the pounded powder of ☿ and the salt of the silver
into a glass *ampulla* or philosopher's egg, and pour on it
the virgin's milk, which, in the previous work, I did not
wish to have mixed with the air, fire and earth. Seal it
hermetice, and set it *in tripode* for 60 days, with a
moderate fire. Then it will change into a hard,
crystalline Stone.

Now break your ampulla or egg, pound your Stone intangibly, put it into a fixing—glass, and pour 4 times the weight of your matter of our white, burning, blessed water on it. Seal it *hermetice*, place it on the furnace in a dish with strained ashes, give it a moderate fire, and everything will dissolve into pure water, also rise and fall, and one will take the other up with it in the fixing-glass; and it will again fall down drop by drop on the matter. They will congeal gradually and one will keep the other with it at the bottom; and one will congeal with the other and stay at the bottom as a fixed oil, clear and transparent, a true Elixir that *tranfers* tin and quicksilver, yes, also gold, into real silver. As *Morienus* says: Whoever cannot make silver of gold, cannot make gold of silver.

Now remove it from the fixing—glass, put it into an ampulla or philosophical egg, seal it hermetice, and set it in tripode for 40 days and nights, or 6 weeks, with a moderate fire, as has been mentioned before. Then it will congeal into the philosopher's stone. Break the glass open, and you will have a snow white stone, as clear as crystal and fusible like wax (or: meltable), which can be pulverized, and which also transfers tin and ☿, yes, also ☉ into true silver, better than that which comes out of the mines. You can at all times multiply this Stone with sublimated ☿, that is, 7 parts of the sublimate to 1 part of the Stone, mixed together, put in a fixing—glass and set on warm ashes. Then let it dissolve in cold water, and rise and fall, till one congeals with the other in a clear, glistening oil, transparent like a crystal, becoming a true Elixir. Then remove it from the fixing-glass, put it into a philosophical egg, seal it hermetice, and set it for 40 days and nights in tripode, and it will congeal into the Philosopher's Stone, which is as good as the first.

CHAPTER LXXVII
THE FIGURE OF THE FIXING-GLASS - No. 24

My child must know that this is the fixing-glass (or: flask) in which all things are congealed. It must be made of thick glass, and its round head must be turned in above, just as the belly of a drinking glass in turned in above. There must be a long, sharp point hanging down, at which point the drops that steam up from below can again fall down. There has to be a tube at the side, to pour the matter and to let the chalk fall in, after which to pour the water in over it. A small round glass has to be cut upon it which fits the tube and closes it well

Now grind some glass-powder so subtly with linseed oil that one could paint with it. Put a bit of that on the edge of the tubes with a small brush, and then put the small glass on it, and upon that put a weight of lead. When it is dry, it is as firm as if it were glass, and no spirits can penetrate through it. After this, you can paint the jar all around with the brush. Let it dry well. You will find no deficiency in your *lutum*, but you will find it difficult to open it up. That you must do with a razor, because this powder of glass and linseed oil dries so tightly together in this way as if it were

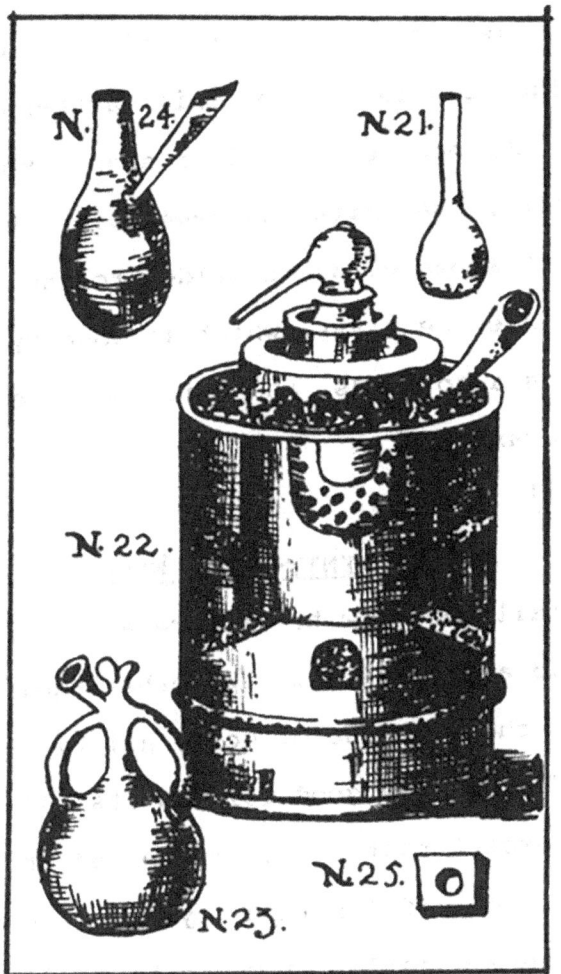

all glass. It stands up to fire, provided it is not burnt; it also stands up to water, provided it does not stand in boiling water day and night. Yet no steam or cold water can harm this lutum. Therefore, lute with this lutum all your glasses, for when I say: "sealed hermetice", I mainly refer to this lutum. But if you wished to burn it in the fire, it would burn away after a while.

Should you have to make a glass burning hot, you must pinch it together with tongs while glowing hot, so that it becomes one glass. There is yet another way to lute with *imitation jewel*, with which the goldsmiths work. This imitation jewel has to be pounded very finely into powder (very gently). One has to mix some borax with it. When one is ready to lute, one has to grind (or: polish) the mouth of the glass quite evenly, and put on it another glass. Then one has to apply this powder with the borax, mixed with some water, over the rim of the glass; the polished glass No. 25 (p. 75) is put on top of it, and a small fire is made above it. The mouth of the glass is put through a tile in which a hole has been made. The fire is made on this earthenware (or: clay) tile, so that the neck of the glass starts to glow well. As soon as it glows, the imitation jewel with the borax begins to melt, and thus one glass is welded to the other, so that both together are one glass, which is done by the imitation jewel, which is also glass, and which melts easily with the help of the borax.

CHAPTER LXXVIII
THE TWENTY-FIRST WORK

My son shall know how he can prepare the Philosopher's Stone from lead alone, without any other additions. It is as powerful as the one made of gold, all by itself, without any ferment, for lead is good gold in its innermost, and it lacks nothing except that it is impure and its innermost is not turned outside. If its impurity were taken from it and its innermost were outside, it would be good gold. All philosophers concur in this who have investigated the work and have found the truth. For all these reasons no other ferment has to be added than ♄ in order to make the Stone.

My child, take as much lead—dross, *mini*, or *cerussa*, 10 or 12 lbs, less or more, as you wish. Take a large, earthenware jar and put 2 parts of strong wine vinegar in it. Add also good, calcinated tartar which has been purified 3 or 4 times from its fecibus, and congealed, burnt in a fire, there upon again dissolved. Repeat that till it leaves no more feces. When it is thus dissolved and again congealed, the tartar is prepared.

Take of that what is required according to how much lead-chalk you have; take also good, clear, transparent ✳, as much as is needed. Put them together in a jar and close them immediately after pouring the vinegar on it, or everything would run out of the jar over the neck. Set it to putrefy into the balneum, and proceed in the same way as I have taught you concerning the Qu. Ess. or ☿ of silver; and in the same manner draw off the Qu. Ess. or ☿ of the lead—calx, mini, or white lead (or: ceruse). (Note: This is lead acetate. HWN)

CHAPTER LXXIX

After you have drawn 5 or 6 lbs of ☿ from the lead, sublimate it in every way as I taught you in the 19th work. Then dissolve it in our *aqua philosophorum* as often (as taught before), and sublimate it again, just as has been indicated in the 19th work of the ☿. When everything has been done and the elements are separated,

137

each in a separate glass, well stopper, and you have kept
the earth at the bottom of the distillation pot of which
you have drawn the fire, take it out of the jar in which
the vinegar, ammoniac and tartar were, since these 4 are
still bodily together in the jar. I have taught you the
operation of the salt, when I instructed you how to bring
the ☿ out of the salt, how to dissolve the matter and
beat down the corpus, and how to dry it. Thus proceed
also with the body of the lead, in all details just as I
taught you then. When you have prepared the salt-corpus,
or earth of the ☿, prepare the salt of the lead, which
was left in the distillation pot, in a similar way as you
have previously prepared the salt of ☿.

CHAPTER LXXX

Now take these two earths or *salia*, that is, the one
that stayed in the jar when you drew over the ☿ and the
salt that stayed in the distillation pot. Take these two,
well prepared as before, mix them well. Take the element
of the air and the element of the fire, put them together
into a glass ampulla or egg, seal them hermetice, set
them for 40 days and nights in tripode, and they will be
congealed into a red, transparent, crystalline stone.
Powder this stone finely, put it into a fixing-glass, and
pour on it some of our blessed, burning water to the Red,
2 or 3 hands above the matter — for which there is no

138

exact measure, since all water is Elixir when it is
congealed with the work. Close it hermetice, set it in
ashes, and it will dissolve in pure water, and rise and
fall together. The light will draw the heavy; and the
heavy, the light; and one will congeal with the other.
But when that happens, all the colors of the world
manifest, more than one can imagine. Then one congeals
with another and stays as a fixed oil, and it is truly an
Elixir.

Now remove your Elixir, put it into a glass ampulla
or egg, seal it hermetice, and set it for 50 days and
nights in tripode. Then it will congeal into the
Philosopher's Stone, transparent and blood-red, and it
transmutes lead, silver, copper, and quicksilver into
fine gold, better than that which comes out of the mines.
No ferment is added to this stone except its own salt or
earth; yet it is as good in its projection and goes as

far as the Stone from ☿ and ☉. The degree of the
projection, however, will be taught to you by experience;
therefore, give thanks to God and be charitable toward
the poor.

CHAPTER LXXXI

My child, you well know that when you conjoined the

work of ☿ from ♃, you took no more than the earth, air
and fire. You kept the water or virgin's milk, because it
did not serve the Red Stone, as has been taught before.

139

But draw ☿ from tin, as has been taught before
concerning ☽ and lead, and prepare the salt as you
prepare the salt of silver. Sublimate ☿ and proceed in
everything as you did with the salt and ☿ of silver, into
which you imbibed the water or virgin's milk which was
left over from the work of ☿. Do likewise in everything
in this work, and when it is perfected, the work of tin
is as good as the work of silver; both are Stones to the
White, and their projection is equally great. That is why
Alcanus says: You must not open your purse to incur great
expense in this Art, because a poor man possesses the
Stone as well as a rich man; he finds it everywhere to
buy, and it is a work of women and a child's play.

CHAPTER LXXXII

THE TWENTY-SECOND WORK, HOW TO MAKE

GOLD FROM ⊕ AND THE RED PHILOSOPHICAL WATER

My child, I will now teach you how to make the Stone
which God gives us for nothing. Know that it can be
prepared in many different ways, but I will now teach how
to make it the way I learned from my father.

You must know, my child, that there are 2 kinds of
vitriol. One is light—green and comes in small pieces;
then there is also dark—green vitriol, and that feels

candied like sugar candy. These two, however, are of no use to us. They are drawn from the top layer of the mines and boiled; they are crystallized like sugar candy. Then there exists still another kind of vitriol. That one comes in big pieces and on the outside looks as if it were rusty iron, somewhat whitish. As the pieces are broken up, they are blue inside, somewhat greenish as if it were a sapphire. These pieces are as big as human heads, weighing about 6 or 8 lbs. Of such you must take, and so as to be surer, buy 1 or 2 lbs of it, dry it at a fire of its wateriness, and then set it 1 or 2 days and nights to burn in a furnace. If it does not become beautifully red, it is no good; but if it becomes nicely red, it is good.

Of this Stone, which God gives us for nothing, take as much as is necessary, dissolve it in distilled wine vinegar and let its feces drop; filter it carefully of its sediment and draw the vinegar off, distilling with a gentle fire till it drips no more and the matter is dry. Then dissolve it again in fresh distilled vinegar, let the feces settle, or draw it off from its fecibus through the felt. Do this 4 times till no more feces remain in the dissolution. Now distill the vinegar off with a gentle fire, till it is so dry as to dust, but nevertheless retains its spirits. Now it is ready for calcination.

CHAPTER LXXXIII

My child must know that this matter is by nature stoppering (or: stuffing, or: constipating) and attracting,
and partly volatile and partly fixed. That is why it has
to be dissolved in distilled wine vinegar, so that it may
retain the subtle spirit of the vinegar, in order to be
calcinated with that spirit, to render it more subtle and
easier to open and dissolve; for the spirit of vinegar
helps to dissolve all things well.

Put this vitriol, thus prepared, into a glass ampulla
or egg, lute it hermetice, but fill it completely, so
that the spirits do not rise. Set it in tripode and let
it stand there in moderate heat, so as to sublimate. Then
remove it, powder the matter finely, put it into a
distillation pot, put a helm on, set it in the balneum
with a recipient, and distill everything that can be
distilled. It will probably be distilling for 20 or 25
days. When it no longer distills, take it off; take the
feces that are at the bottom and pound them intangibly on
a stone. Now put them into a dissolving glass (or:
flask), and pour all the water upon it. Seal it hermetice
and put it into a hot balneum. Then it will dissolve into
pure water without settled feces. Remove it and pour it
into the distillation pot; put a helm on, and again
distill off all water through the balneum with a moderate
fire. Preserve the distilled water well. Put the pot with
the helm on a furnace in a dish with strained ashes, a
recipient on it, and receive the element of the air as a
noble, yellow oil. This is done with strong heat. Keep it

and stopper it and put it next to the water, each by itself.

Now take the feces from the distillation pot. They are red as blood; pound them intangibly on a stone, put them into a glass ampulla, or egg, seal it and put it to subtilize for 30 days and nights in tripode with a moderate fire. Afterwards, remove them, pound them on a stone, put them into the dissolving glass, and pour over it all the element of the water; seal it and put it again in the balneum, as before, and let everything dissolve into pure water as before; remove it and pour it into your distillation pot, put a helm on, set it in a dish with ashes on the furnace. Put the recipient into cold water with its belly; first, give little fire, increasing it gradually till it glows strongly, and let it stand in this heat for 5 or 6 days. During this time the element of the fire will go over in the form of a red oil. Let it cool down for 3 or 4 nights. Then remove the recipient, close it tightly and put it next to the element of the air and the water, till you need it.

CHAPTER LXXXIV

Now remove the earth or feces which stayed at the bottom and look greyish-white, provided the element of the fire has been well drawn out. Pound them small on a stone, and put them to calcinate in a reverberation—furnace for 3 days and nights, with a moderate fire, so that they glow nicely and no more, as has been taught above. Then remove them, put them into a stone jar, pour distilled wine vinegar on them, stir well with a wooden spoon so that they get well mixed. Then put them, well stoppered, into a boiling balneum for 4 days and nights, meanwhile stirring the matter into the vinegar 5 or 6 times a day, each time closing the jar again. Finally, let it cool down and let the feces drop; pour the vinegar carefully from the fecibus, and filter it; dry the feces, so as to use them in *chirurgie* (surgery); pour the vinegar into the distillation pot, put a helm on and a recipient thereon, draw the vinegar off in balneo, and the element of the earth will stay at the bottom snow white.

Now set it again for 24 hours to *reverberate*, in moderate heat and no more. Remove it and put it back into the jar, pour vinegar on it, put it in the balneum as before, again distill the vinegar off, and your earth or salt is prepared. Pound it intangibly, put it into a glass ampulla or egg, pour on it all your fire or air; but do not use the water; use it in medicine, and seal it hermetice, put it in tripode for 40 days and nights, and it will congeal into a Red Stone.

Remove that and add to 1 lb of it 2 ounces of gold oxide, and 2 lbs of our red, blessed water. Put it into a fixing-glass, seal it, and let it stand on warm ashes till the gold oxide is dissolved into pure water. Now rub your aforementioned Stone to a fine powder and add it to the calcinated and dissolved gold in the fixing-glass; seal it again and put the glass on the furnace in a dish with ashes. Thus, by means of the regulation of the fire,

144

everything will be dissolved into pure water and rise and fall together. Finally, most of the colors will manifest, more than one can imagine. At last, it will turn into a red color and fixation, staying at the bottom somewhat thicker than honey. Do not let it cool down, as otherwise it would not run out, but while still warm, let it run into a glass ampulla. Seal that hermetice, and set it for 6 weeks in tripode with the same regulation of the heat as before. Now it will congeal into the Philosopher's Stone. Remove that, thank God, and be charitable to the poor, for you have a precious treasure, which transfers silver, lead, copper,: and quicksilver into true gold, and this Stone can be multiplied as has been taught above.

CHAPTER LXXXV

THE TWENTY-THIRD WORK OF ☉ AND ☿

My child, I will now teach you how to make the Stone from ☿ and ☉. Take quicksilver, as much as is needed, sublimate it to the Red; after that, dissolve it in aqua philosophorum to the Red, as has been taught above in the work of ☿. Bring it to calcination, then take ⅔ ij gold calx (oxide) to each lb of the thus prepared. Pound it very gently on a stone together with the ☿, put it to sublimate; sublimate ☿ through the ☉ oxide ("Kalk"

145

actually means "chalk" or "lime", but Goldkalk is
translated as gold calx or "gold oxide"). Again stir what
has been sublimated into its oxide or feces, and
sublimate it again, a third time. Finally, keep your

feces of the gold, and again take $\mathsf{3}$ ij gold oxide;

pound it again with the same sublimated $\math{\displaystyle \mathop{\text{☿}}}$ 3 times, as
before, repeating each time with fresh gold oxide; and
keep all your feces from the gold together. Put them in a
reverberation furnace for 8 days and nights, to calcinate
in moderate heat, as before. Then your matter or feces or
earth of the gold is well calcinated.

Now take our burning blessed water to the Red, 2 lbs;
put it into a fixing-glass and add your matter or earth
of the gold, seal hermetice and put it on the furnace in
the dish with the ashes. Thus it will, given moderate
heat, dissolve into clear water. Pound your sublimated
quicksilver very fine on a stone, put it into a glass
ampulla, set it to calcinate for 40 days; after this,
take it out, pound it and put it into a distillation pot,
add a helm with a recipient, set it in balneo to distill,
and distill everything that can be distilled. But you
must distill it for 20 or 25 days and nights. When it no
longer distills, take it out and pound it on a stone. Put
it into the dissolving glass, pour all your water on it,
which you had drawn off it; seal it hermetice, put it in
the balneum, and everything will be together dissolved
into pure water. Proceed in this work in everything as

has been taught above in the work of ☿, when you separated the elements.

When this has been done and your salt of ☿ has been prepared, pound this salt intangibly on a stone, put it into the fixing-glass together with the dissolved gold, and add all your air and all your fire, but keep the water or virgin's milk which does serve no useful purpose for this work. Seal the glass and put it on the furnace in a dish with warm ashes. Now everything will dissolve together and also rise and fall together, till all colors appear and they finally turn into a red color and fixation, staying at the bottom like a fixed oil. While still warm, pour it out of the fixing-glass into a glass ampulla, seal it and set it in tripode for 60 days and nights, with a moderate fire. Then it will congeal into the Philosopher's Stone. Take that out and thank God. This Stone exceeds the virtue of all Stones which have been taught so far, because the ferment spiritualizes, and its salt is prepared.

CHAPTER LXXXVI

THE TWENTY-FOURTH WORK OF ARSENICUM

Now I will teach my child how to make the white Stone from arsenic. Take arsenic, 4 or 5 lbs; powder it finely; then take alum, egg chalk (or: egg lime), calamine and common salt, all dried at a gentle fire, that is, 2 parts of common salt to 1 part of each of the others. Pound them (or: mix them) well together, and for every 4 lbs of arsenic, take 8 lbs of the other matters. Mix them together and put them into a sublimation vessel; sublimate the arsenic. Pound what has been sublimated among its feces, and sublimate it again. Repeat it once more. Now mix the *arsenicum sublimatum* with as much fresh matter as your arsenicum weighs; sublimate it and repeat 3 times. After this, sublimate again 3 times with as much fresh matter. Then your arsenic will become clear, white and transparent like crystal. Dissolve that in *aqua fort*, made of alum, egg chalk, calamine and saltpetre, as much as all the others weigh together. Dry them to the point of dusting, put them into a distillation pot and distill as one normally makes aqua fort. Put the *caput mortuum*, powdered, back into the pot, pour your aqua fort again on it, distill for 36 hours till nothing drips any longer. After this, keep it glowing for another 36 hours; repeat that again, rectifying it with its caput mortuum. It is achieved with the third distillation.

With this dissolve your sublimated arsenic, draw the water off, and sublimate the arsenic again; and again dissolve it in the said water. Do this 3 times, each time dissolving in fresh water. Then sublimate the arsenic by itself, and it is prepared.

CHAPTER LXXXVII

Now take as much silver as the weight of the arsenic, dissolve it in common aqua fort, and beat it down. Wash this chalk with common water of its saltiness, and dry it on a moderate fire. Take this *calcem Lunac* and the

148

prepared arsenic, pound them together on a stone, put them into our *sublimatorium*, and sublimate them of the chalk, at first heating fairly strongly, so that the distillation pot stands there in a soft glow, for 2 hours. Then let it cool down, and remove the sublimate; again pound it with its fecibus, and set it again to sublimate as before. Do this 4 times, and you have sublimated all the spirit or Qu. Ess. of the silver with the arsenic. Then put the feces of the silver to *reverberate* or calcinate in the sublimation furnace, for 4 days and nights, in a soft glow not too hot or it would melt and spoil everything.

Then take it out, put it into your stone jar, pour distilled wine vinegar on it, set it in the balneum for 4 days and nights, and proceed in every way as I have taught above in the work of *antimonium*, when I instructed you how to prepare the salt, earth, or corpus. Subsequently, pound it together intangibly on a stone, and for that take twice as heavy of our burning blessed water to the White. Put everything together into a fixing-glass, seal it hermetice, put it on a furnace in a dish with ashes, and again give moderate fire, and everything will dissolve into pure water, also rise and fall, until it stays fixed as a crystalline, clear oil,

which is a perfect Elixir. It translates ☿ and ♃ into true silver. Put this into a glass ampulla or egg, seal it, set it for 40 days and nights in tripode, and it will coagulate into the Philosopher's Stone, whose projection is very great. The very same work which you have here

149

done with silver you can also do with tin, and the projection will be equally high. Thank God and be charitable to the poor.

CHAPTER LXXXVIII
THE TWENTY-FIFTH AND LAST
WORK OF THE SULPHUR

My child, listen and hear about the strong Stone above all Stones, as well as the greatest secret among all those taught before. My child must know that there are different kinds of sulphur. There is blackish sulphur (or: darkish) which comes out of iron ores. There is green and yellow sulphur, which comes from the ores of ♀. There is sulphur which is somewhat yellow and greenish; that one comes from the ores of lead. There is also white sulphur like suet, which comes from the ores of ♃ and all these do not serve our work.

But there exists a *Sulphur* which is hard and schistous and somewhat clear, and very beautifully lemon—yellow, as though it were noble gold. This one occurs in large lumps of 10 or 12 lbs. When it is broken up into pieces, it is very light yellow inside, threaded with small streaks, and somewhat reddish. If one knocks off a piece of it and rubs both together, it creaks and whistles like little mice; or if you rub it with the nail of your finger or a stiff leather, it creaks. Yet a better test: Take a flat piece of ☽, rub it firmly upon it; or take powder of this sulphur and put it on a silver platter, and rub it firmly with a dry leather, the ☽ will take on a beautiful yellow-red color, as if it were polished copper. That one (that Sulphur) is good. Or put the Pfennig (a German coin) or silver for 8 or 9 days, thus rubbed, in a dry spot out of the air. If now the Pfennig becomes black, the *Sulphur* is no good; but if it stays beautifully red like polished copper, it is good Sulphur. You find this Sulphur in Frankfort, and it comes from Hungary. One can also find it in Cologne, or in Middelburg and Brugge (Note: In French it is "Bruges"); that one comes from Spain from the gold mines, and it is the right one.

Of this Sulphur take 10 or 12 lbs, pulverize and pound it on a stone with distilled vinegar, as if one were to paint with it. Put it into a stone jar and pour on it a goodly quantity of ✳. Set it in the *Baln. Mar.*, stir it well together with a wooden spatula, cover the jar, and let the balneum boil for 6 or 8 days, stirring every day 3 or 4 times; then let it cool down and fall. Filter the tinged vinegar off and pour other vinegar on. Do this till the vinegar begins to become colored. Put all of the colored vinegar into the balneum, and distill it till there is a golden—yellow powder at the bottom. Rub this powder again with fresh vinegar on the stone, put it into the balneum with ✳, as before, and draw it off till no more feces remain in dissolving it. Now distill the vinegar off. As to the feces that remain, you can collect them together and melt them and sell them to the manufacturers of gunpowder.

Weigh the powder, and to 1 part of sulphur-powder take 6 parts of Roman vitriol, which should be dried Ⓑ , but in such a way that it keeps its spirits. Pound all this together on a stone, put it into a *sublimatorium*; sublimate it first with a gentle fire, and then *gradatim* (gradually) more, and finally strongly, and do that (strong fire) for 2 days. Take the sublimate out, mix it with its feces and sublimate it again. Do this 3 times. Then throw the feces away, put it back with fresh matter, as has been said, and mix it 3 times with the feces; and then again with fresh matter, 3 times; then dissolve it in our red philosophical water, which has been taught here. Draw the water off and sublimate it 3 times.

Now weigh your sublimate, and take to 7 parts of it 1 part of powdered gold oxide, then it is ready. Now take your sublimate and put it into a golden philosophical egg, because the glass is of no use here as it would become soft like wax. Close it tightly, and set it in tripode for 18 weeks, the first 6 weeks as has been taught in the previous work; the other 6 weeks such that one could light a paper if it were held in the jar in which is the egg. Try that every day 2 or 3 times or more, and block the square hole at the outside furnace in order to better retain the heat. Let it stand in this regimen till the 18 weeks are over.

Meanwhile, however, while the work is standing in tri-pode, calcinate your *feces Solis* which remained after the sublimation. Reverberate the *calx Solis* for 20 days; put it into a fixing-glass, pour on it our red burning water, just enough that it can dissolve. When it is dissolved, pour it into a small glass, close it well and keep it for a time of need. (Note: When the golden cask has been used, it no longer serves this work to make medicine from it, for the matter has drawn out the spirit as being its like. But it is just as good to sell it or the like, it is only a little lighter and paler.)

CHAPTER LXXXIX

When the 18 weeks are over, remove your matter and take as much of it as you wish. Powder it, put it into a fixing-glass, and pour red burning water on it, as heavy as the powder weighed. Seal it or close it very tightly; let it dissolve, clarify, take it out, and pour it into a distillation pot with an alembic. Distill it per balneum in clear heat. The recipient has to be very well luted. You must have an alembic which has a tube above, because you must distill 6 times, each time pouring fresh red water on; and your matter will remain in the glass like thick honey. Then remove it from the balneum, set it on the furnace in strained ashes, first give a gentle fire, and subsequently *gradatim*. First the air will go over as water.

After this, one has to put on another recipient, and it will go over as a gold-oil, so that the helm and the recipient will shine like gold. Let it stand thus till the helm will take the color of blood. Now remove your recipient and close it quickly; put another on the spout, for 24 hours, increasing the fire *gradatim* till the pot starts glowing at the bottom. Then it will go over like blood and finally like red smoke.

When no more *spiritus* go over, let it cool down, take it off, and set the feces to reverberate for 8 or 10 days. Then take them out and put them into a small glass ampulla. Add the solution of the gold, which you had kept in a small glass, to the other earth in the ampulla. Now you have both earths together. Put them on hot ashes and the earths will dissolve together. Then set it in the balneum and draw the water off, and a salt or earth will remain of your matter which you must pound and put into a glass ampulla or egg. Pour on it all the air with all the fire, seal it hermetice, and set it in tripode for 12 days, with a moderate fire. In time all fire and air will be imbibed into the earth and become hard as a stone.

Remove that and weigh your matter; put it into the fixing—glass and pour on it our burning water, as much as the matter weighs. Seal it hermetice, and set it on the furnace in a dish with ashes, giving a moderate fire. Now everything will dissolve into clear water, also rise and descend and one will rise with the other and subsequently, one will also keep the other at the bottom and make it fixed. But before all that takes place, all imaginable colors will appear; and when that is over, everything will turn into a white color. Do not increase

154

your fire till you see that the matter turns pale yellow. Then rejoice: For under the White the Red is hidden.

Now increase your fire considerably, and take good care that it does not cool off before you see your matter perfectly red. When it is red, pour it into the egg while rather hot, and seal it with the seal *Hermetis* (the seal of Hermes). Set it in tripode with a good regulation of a small fire, so that you could easily keep your hand between the wall of the big iron furnace and the chamber in which stands the egg with the matter. Within these 6 weeks it will coagulate into a Philosopher's Stone.

Take it out, and thank God, for it is the Stone of which much has been said before. Its projection is infinite, and it makes the best gold that can be seen, though it is all too malleable; so one has to add to this gold other metal-gold, because it is all too soft.

END OF THE THIRD PART OF THE MINERAL WORK,
ISAACI HOLLANDI

JOHANNIS ISAACI HOLLANDI

CHYMICAL TRACTATE

CALLED BY HIM

THE HAND OF THE PHILOSOPHERS

With its Secret Signs

PREFACE OF THE TRANSLATOR

After I had recently published the valuable third part of *Opera Mineralis Hollandi*, whose finding pleased me greatly and even more so as it is now in print and can no longer be withdrawn from the lovers (of the Art), and because, in my opinion, the present tractate deserves no less consideration, I could not — because I received those writings all together — rest till I had translated them from Dutch into our High German language, so as to get them also into the hands of my co—lovers, the usefulness of which will become apparent of itself after diligent reading. Do not doubt, therefore. We shall now have most of the writings of this very dear man in print in Latin and in German, that is, those which are the most important. Yet it would be good if the first two parts of *Opera Mineralis* were likewise accurately translated into High German from a true Dutch copy and not from the Latin text (because I have no time to do it). For it is possible that the Latin version cannot be wholly trusted, seeing that the same translator of the *Opus Vini* has not done his job too well. Even if the Latin text were correct, we have so many High German compatriots who do not know Latin but who also deserve to know it (The Hand of the Philosophers).

If someone were to object that, although it requires a great deal of effort and knowledge, there is not so much to the writings of Hollandus, a man who would judge in that way would reveal his crude lack of intelligence, since one can learn more from Hollandus than many a man can imagine. I am of the opinion that if all other chemical books were done away with, except *Basilius Valentinus* and *Paracelsus*, there would remain enough to learn from them alone. Yes, many a man who had botched and bungled on wrong ways for many years, when he finally comes across the books of Hollandus, finds the mistakes he made in the past vividly depicted, yes, perhaps even his future mistakes which he would have continued to make if he had not found better instruction in *Hollandus*. Reading his printed books has so far not been as fruitful as it will henceforth hopefully be, because at <u>that</u> time alchemists did not understand his furnaces and instruments to which he refers everywhere. However, these have now been sufficiently described and illustrated in the third part of *Opera Mineralis*, so that I hope that now, due to my diligence and promotion, more progress can be made in *Hollandus* than before.

Yet, I am aware of one obstacle in this author's writings which may deter many a reader. It consists in the fact that the author has the fault of repeating himself so often that many have found it more of a nuisance reading those repetitions than the author did in writing them. If such tautologies were cut out from the author's writings, many would have more pleasure reading them, as I myself must admit that I found it a great nuisance translating that type of thing. But the reader must know that there are all kinds of people; consequently, also those to whom one cannot tell a thing often enough. What is said too much to lovers of brevity, therefore, is still too little for them.

For their sake we have to let the said tautology pass. Even without that, however, it is not up to us to change anything in the writings of famous people. The praise that I rightly accord to the writings of our author and whose reading I recommend should not be interpreted to mean that I am seeking to persuade every coarse reveller to plump with unwashed hands into these very important writings and to copy the Art without discernment. This is not my intention at all. I rather predict a totally fruitless effort to those who would do that, since in many things Hollandus wants to be understood quite literally. To do this, therefore, a trained reasoning mind is required if one is to benefit, in view of the fact that for teaching handy masters and simultaneously unhandy ones, just this *Hollandus* is as accomplished a master as can be found in the whole *Turba Philosophorum*.

Consequently, it is a pity that all his *Opera* as he wrote them one after another, are not put together into one *Corpus*, as *Arnaldus De Villanova* and other *Opera* are available. For, it may be assumed that what he wrote was not only all *Chymica* but also much *Medicina*, because he had been an excellent *Medicus*. God only knows where all his writings have been dispersed, because even at this time no one can be found who could give the slightest information on the place where he resided and at what time he actually lived. (Editor's Note: See Appendix A)

Yes, we cannot even distinguish the works in print as to which were written by *Hollandus Pater* and which by *Hollandus Filius*, although that both wrote them may be seen in different places; also that *Hollandus Pater*, yes, also his grandparents, excelled in *Arte Chemica*, just as it is mentioned in *Opere Saturni* that *Hollandus'* grandfather was the inventor of *Olei Plumbi Philosophorum* (Note: Olei is the Genetive case of the oil; nominative is: oleum), which had not been known to the old people. This leads us to believe that his grandfather lived at the time of *Raimundus Luilius*, who wrote at least a hundred years before *Hollandus Filius*; because *Lullius*, in his letter to King *Rupertus*, also mentions that *Oleum Plumbi Philosophorum* was a new discovery and such a highly important secret that it seemed unbelievable to all old alchemists, because with it the *Lapis Philosophorum* could be wholly perfected within thirty days after its first fixation.

From this we conclude that *Hollandus Filius* had not become such a great master without reason, for he had as it were, inherited *Arcana* from his whole *Familia*. None of those who quote from his writings take note of this, but

159

each writes *Johannes Isaacus Hollandus*, which, however, is mere confusion in my opinion. It should be written *Johannes Isaaci Hollandus*, because the Dutch give only one name to their children. Neither do their children have a second name, but the father's first or Christian name with the addition of the word "sen" or "son". Consequently, according to Netherlands custom, the Younger Hollandus was called *Jan Isaacsen*, that is, Johannes son of Isaac. To mention this here does not seem to me to be inappropriate.

Some other tractates are ascribed to our author which are supposed to be still extant, such as, *De Oleo Stibbi*. It is no doubt the tractate which is in *Basil's Triumphant Chariot* under the name of Baconis (Note: authored by Bacon). Yet by the style one may well infer that it is by Hollandus: *Item Secretissima Revelatio Manualie Operationis Lapidis Philosophorum., Item De Sulphuribus &.* These, however, I have not yet seen. If I obtain any of them, I will not withhold it from the lovers of Alchemy.

On the other hand, I also beg and exhort others, if they obtain a good writing, to apply the same measure to me as I apply to them; that is, to allow it also to be printed, so that we lovers get an opportunity to read it, in consideration of the fact that those works were written by their authors with the intention of promoting much good thereby. If then the good is to be promoted, it must get among the people. In a box or locked bookcase it lies immobilized, and nothing comes of it unless it gets into human hands and is read, so that some improvement in teaching, life, or manual operation may be achieved through it. It is difficult enough for it to bear fruit, and no hiding of good books is necessary; for as it is, they are hardly useful, even if the alchemists do all within their power and must first themselves surmount various difficulties. Thus, for instance, to write a good book is difficult to begin with, unless there is someone who has received the gift from God to write thoroughly; and when it has been written with a great deal of talent, it is difficult again. Because of envy, dogs in the manger like you and your like may get a hold of it, lock it away and keep it imprisoned for the rest of their lives, so that it neither benefits themselves nor others.

When it has also escaped from this danger, it again has difficulties in reaching such men as are able to understand it. And when there are some who do understand it, it is again difficult to come across the book. If they do come across it, there is danger that the devil might prevent their getting the book. Finally, when these obstacles have been overcome, it is questionable whether it will help some people because of their unworthiness.

How should we powerless men not fare thus in our work for the good, since the Almighty Lord Jesus has fared the same way: All his labors are benefiting only the least number of people. Accordingly, as I mentioned, also good books that are not suppressed cannot accomplish much good, let alone that they are at first hidden and are rather granted to cockroaches and worms, which are curious enough without that.

On the other hand, there is an excess of useless books bred every year like harmful vermin ("under changed titles, but always with the old tune"). Thus wrote *Hippocrates*, *Calenus*, *Lutherus*, *Augustinus*, the *Corpus Juris*, etc. — as if one could not see it oneself and first required a thousand foolish coaches and had, in addition, to render homage to them before great *Doctores* and *Licentiatos*, on account of their lousy citing and hundred-fold copying of other people's books. The majority of their authors stand in need of first becoming good disciples and of learning some honest trade. But most of them reverse this and soon rise from their childish youth to the Doctor or highest teacher degree. Because they cannot earn it by the right kind   of talents, they graduate through money, just as if one let common water pass for wonderful wine through a written false certificate.

This is exceedingly strange to hear, and yet it is quite the fashion, not only in the world but even in Christendom, to graduate through money such young fellows who were still in grade school hardly a few years before. Yes, sometimes not so much time has lapsed since their neighbors saw them ride on sticks with other boys, or play other childish games. From where then should they get the great art of acting as *Doctores* and of writing extensive volumes, for which by right, a special gift of God and many years of experience are required.

Nevertheless, such books without pith and power are written in heaps. Would God, that they were withheld and instead those of the dear men, of whom one is hardly seen in fifty or a hundred years, were strongly promoted. How long has it been since *Paraceleus* lived, and we still have not got his *Herbarium*! Likewise, we are still missing most of the dear *Carrichter*. Those who deliberately withhold those are public world-thieves, be they who they may. But enough of this, and I hereby recommend the reader to God's protection and the inspiration of the Holy Ghost.

Written in February 1667.

Hand of the Philosophers

This is the Hand of the Philosophers with their dear secret signs, with which the old sages united with each other and took oaths. Nobody can understand this Hand with its secret signs, unless he becomes first a juror of the philosophers, (one who swore loyalty to a philosopher), and has loyally served them in the Art Alchemia. Consequently, those who have not this Hand and do not understand its secret signs, nor have taken the oath of loyalty, are bastards in this Art. They do not possess the philosopher's treasure. That is why I advise all those who do not possess the secrets of the Hand not to start working in the alchemical Art, nor to believe books or writings, since they will all only be cheated in the secrets of the Hand. Therefore, everybody had better be careful.

In this Hand is locked the secret of the philosophers, that is, of the seed and the earth, as will be told later.

Now then, I will teach my child and describe the secret, hidden matter of the wise philosophers and masters of the true Art Alchemia. Nobody can use it unless he take the oath and swear not to divulge the Art and secrets and hidden signs of the sages, except he finds that it would be a good placement. In that case, he should also request the oath that that man should not use the Art except for the salvation of his soul. Only then can he be given the secret signs of the philosophers or sages, with their hidden signs and meanings.

1. <u>THE THUMB</u>

First look at the thumb on which stands the crown next to the moon, one quarter old. By this is meant saltpetre. For just as the thumb vigorously finishes off the hand, saltpetre does in the Art Alchemia, for he is the King and Lord of all salts. He is the mill through which everything must be ground. His nature is elsewhere sufficiently described.

2. <u>THE INDEX</u>

The second sign and secret of the philosophers is the STAR with six points, standing above the foremost finger next to the thumb. It is compared to Roman *Vitriolo*, because no work that is to be perfect can be completed without vitriol; for it is the greatest and strongest salt after saltpetre. It's nature is described.

3. THE MIDDLE FINGER

The third sign of the philosopher's Hand is the SUN, standing above the third finger. By it, *Sal Ammoniacum* is designated, for apart from saltpetre and vitriol, no

thing more powerful is found than ✳. That is why it is the third secret.

4. THE RING FINGER

The fourth sign of the philosophers is the LANTERN, standing above the fourth finger of the Hand, whereby *Alumen Roche* is indicated; for without alum, no perfect work can be accomplished, because it is required for the Red and the White. It has an astonishing nature and the most subtle *Spiritus*. It's described elsewhere.

5. THE LITTLE FINGER

The fifth secret and sign is the KEY of the philosophers, standing above the little finger. Simultaneously, it is the lock of the Hand. That is why the key is standing on it. By it, common salt is designated, for salt is the Key in this Art.

6. **THE MIDDLE OF THE HAND**

The sixth secret sign is the FISH. It stands in the
middle of the Hand and signifies *Mercury*, for without
☿, or the fish, nothing can be done. He is the beginning,
the middle and the end, and he is the priest who must
marry everything. And he is the male and the seed; he is
the water out of which all metals have originated; and he
is the principal (factor) of all Arts, and the greatest
of all secrets.

7. **THE PALM**

The seventh sign of the philosophers is FIRE . By this
Sulphur is indicated. It is the earth and beginning of
all metals. It is the female who brings forth the fruit.
For no seed can grow unless it be first thrown into
fertile soil. Then beautiful fruit will come from it.

Thus it also happens that when a pure ☿ is joined to a
pure △, it brings forth pure fruit. Thus, they are man
and woman, father and mother, fire and water, seed and
earth. This is sufficient about the seven secret signs of
the philosophers. He who understands well this Hand and
its signs, and can work with it, will derive joy from it.
Now follows the *Praeparation*.

THE PREPARATION OF SALTPETRE

Take living chalk (quicklime), according to the quantity of the substance. Pour on it a good amount of urine. Let the lime slake in it; after it is slaked, let it settle, and our it off above (decant).If there are 6 lbs. of saltpetre, take 12 lbs or pints of pure urine, even a little more, but not less. Put it all together into a clean kettle. Boil it and skim it with an iron fish spoon. Put the latter occasionally into the *Liquor* and squirt into the fire. If it burns, or the coal becomes ignited by it, it is enough. Take it off and let it cool somewhat. Now pour it into a large linen sack, like a claret sack. This sack has to hang above a barrel, 5 foot above. Soon there sprout cones in the water, one above the other, as if it were crystal. Take these out - it is the purified saltpetre. Now take the other saltpetre which did not sprout into cones. Fish it out and put it into the urine. Let it boil again as before and pour it also through the sack. It will immediately crystallize into long streaks, like the first. What stays behind is good for nothing; it is only salt which can be coagulated, and then it is common saltpetre.

THIS IS THE MANUAL WORK DONE WITH THE CROWN
OR SALTPETRE

Take 4 lbs of the Crown, prepared egg chalk, *Sal Alcali* made of quicklime, weedashes, potash of *Sal Vitri* of after-wine (residue of wine) or shoots of vine, calcined *Tartarum*, in equal amounts. These things must be prepared with the hand, as you well know. When they are prepared, dissolve them in distilled vinegar. Hang them in the *Balneum* to dissolve there for 21 days and nights. Then coagulate them in the *Alembicum*, and keep the matter clean, so that no dust or impurity can get at it. After this, put them into a round, earthenware vessel, as illustrated here. Put it *In Tripode* or in the *Athanor* till the King is fixed. Regulate the fire so that you can barely keep your hand in it — for the first 8 days. After this, let the vessel glow for another 8 days; but before you do this, divide it as if you wished to burn separating water; drive everything over that will come over, for it is of no use to the work, since the volatile spirits spoil all works, so that they cannot reach the state of fixation. Therefore, before putting it *In Tripode* to calcine, get rid of the volatile spirits, otherwise your work would not get fixed.

When it has been standing *In Tripode* for 14 days, take it out. Take 2 lbs of the Long Finger, which must

have been sublimated 3 times through salt. Pulverize it
with the powder you took out of the Tripode, and
sublimate them together till the Long Finger becomes
fixed.

Now dissolve this fixed powder 7 times in good, distilled vinegar. Let it settle each time, and each time
separate the *Feces*, and congeal it again in the *Alembic*.
Then glow the King in the fire, but take care that he
does not melt. Do this 7 times, by dissolving,
congealing, and glowing as before. Then the saltpetre is
well prepared for the work. Keep it in a closed box of
CYPRIAN earth, glazed with glass, and you have an
infinite treasure, greater than you might believe , with
which you can accomplish wonderful things. I will
describe to you only one part of its effect, should you
need the Art: By it you can turn all seven metals into

their first nature, that is, into ☿ ✝.

To this end you must take a good amount of

distilled vinegar and ⚚, and ⧗ of raw ✳; add to
it whatever metal you wish, provided it is calcined, then
quickly seal the jar and put it in the *Balneum* or in
horse manure for 21 days. After that, put it on fire with
the *Alembic*, and distill. First there will be vinegar and

✳, afterwards the essence of the metal, that is, ☿ ✝.
With that you can do astonishing things. But there is a
great deal of cheating in this; that is why we do not

170

wish to describe it. When you have used the King and he has done his work, let the feces drop, pour off the pure, and coagulate as before. Then he will be better than before, because his power has grown tenfold by having been dissolved and purified with

✛ and ▽, as also next to the ✳ , and by having been water and then congealed. One performs miracles with this King; that is why he carries the crown.

Also, he makes all red metals white, and all soft metals hard, and all hard metals soft, and I write more than I have been commanded to.

Also, make an Aqua fort of ⊕, ◯, ⊕ and ✳ am; and to a quarter lb of this AF., add ろ ij ✳ ; dissolve in it ij lb ☿; draw off the Aqua fort *Per Balneum*, and you will thus kill ☿. Then take 4 lbs of common salt to every 2 lbs of this *Merc*. Mix and pound it well together in a mortar; then sublimate it; then take it out and pound it again with ⊖ as before, and sublimate it. Repeat this 7 times.

After this, take 1 lb ⊖, ½ lb ✳ , mix them well with the sublimated Mercury, and sublimate again. When it is sublimated, take it out, and mix it again with fresh salt alone; sublimate again, and do this 3 or 4 times.

Now pound it fine and put it into a glass vessel and set it to calcinate into the philosophers' stove for 30 days and nights.

Then remove it, and dissolve it in *Balneo* or on a stone in the cellar. When everything has been dissolved, put it to distill 7 times in *Balneo*, till everything has become hard. Then take it out, pulverize it small, and imbibe it with water of ☽ made thus:

Take fine cupel silver; cement it till it comes out white from the ⊖ ; dissolve it in AF.; then draw it off *Per Balneum*; take it out, add to it as much ✳ as there is ☽ , pulverize them together, and dry it. Then calcinate it for 30 days in the philosophers' furnace. After this, dissolve it in *Balneo* or on a stone. Imbibe this water into your powder, and put it into a glass to distill *Per Alembicum*, so as to draw off the humidity.

When it is dry, take it out and imbibe it again with ☽ water. Do this till all the water of ☽ has been imbibed into it. Finally, let it stand on the fire till all your matter is fixed. Then make your matter fusible till it melts like butter.

Also, when you have drawn out the Red from the foremost finger, or from the fire, or from several other things, take 1 lb of the dry powder and 1 lb of the sun.

172

But the sun must first have been 3 times sublimated through the key till it is transparent. Then take 1 part of the Red, ½ a part of the sun, powder them together and sublimate them. Then the sun will sublimate up, and the Red will stay at the bottom of the vessel. Take the sun out above and mix it again with the Red that stayed at the bottom of the vessel. Sublimate it again as before, at least 10 or 12 times. Now take

the same sun and the same Red, powder them impalpably, and put them into the philosopher's furnace to quickly calcine for 30 days. Then take it out and dissolve it in *Balneo*. When everything is dissolved, you have a brilliant water with which gold would not want to be compared. Imbibe your elixir with this red water as you know.

PREPARATION OF VITRIOL

Take 3 or 4 lbs. *Vitriolum Romanum*, dissolve it in *Balneo*, clarify it of its own impurity, and calcine it till it begins getting grey. Then dissolve it again, and let everything become pure again. Then calcine it until it becomes yellow. Now take the *Tincture of Auripigment* and gradually imbibe it into it. Dry it carefully, imbibe it and dry it again until everything has been imbibed into it. Then it is ready to sublimate Mercury through it, so long that he no longer desires anything. Then his stomach is full, and add each time fresh *Species*. Then it turns into a precious ☿ and a salt more splendid than gold.

AN ART OF ARTS

How to extract QUINTAM ESSENTIAM from ⊕ or the
PHILOSOPHERS' STONE
from the Hand.

Take 6 lbs. of glorified vitriol, as you well know how to make. Dry out all its humidity. Then take 3 or 4 lbs. of the Fish and 1 lb. of the prepared Key. Mix them together and put them to sublime. The *Tincture* will rise with the Fish in a secret and invisible way. If then you

wish to sublimate the Fish once again with fresh ⊕ and fresh Key, you may do so. Then you have the fish full of *Tincture*, but if you wished to separate the latter, make a water of the Hand, or a bad water with salt, or made with some of the long (Middle) finger. Put the sublimated Fish, which is full of the *Tincture* of the Star, into the *Balneum*, and it will dissolve into water.

Now take 6, 8, or 10 lbs of the Key, which has been prepared to sublimate through it. Put the Key into a Syburg jar and pour into it, upon the Key, what has been dissolved, so that it turns into a dry *Materia*. Otherwise the water would not putrify or go up in smoke (steam). Put the jar *In Balneum*, add a helm, and distill the AF. over in a boiling *Balneum*, till nothing comes over. Following this, set the jar into sand, and more spirits of the AF. will go over which did not rise out of the Balneum.

When all the *Spiritus* have gone over, increase your fire and you will see the *Quinta Essentia*, or *Tincture* of the Fish and the said Star, sublimate as if it were a crystal stone. Remove it carefully from the *Alembico* and dissolve it again in fresh AF. Repeat this 4 times, and the tinctures of Qu. Ess. of the Fish will become so subtle that it is unbelievable to write about it. The reason for it is that it was first dissolved and afterwards sublimated, and that it is killed and rises into a spiritual *Corpus*. In my estimation, it should be distilled and afterwards sublimated so often that it will obtain such great power that it would be worth a kingdom.

Now you have the Qu. Ess. Vitrioli and ☿ ii together.

If you wish to make an *Elixir* of it, you must calcine it *In Tripode*; then dissolve it *In Balneo* with the water of the Hand; distill and sublimate it, and add the soul to it, and congeal it, as I have taught of other stones. But if you wish to separate the tincture of the Star from the Fish, take good ✳ and distilled vinegar. Pulverize the Quintessentia and put it into the vinegar. Set it into the *Balneum*. The ☿ will sink to the bottom, and the Qu. Ess. of the ♁ will go into the vinegar. Pour off the tinged vinegar, and pour fresh upon it. Put it back into the *Balneum*, and repeat this till the vinegar no

175

longer tinges. Now put all the tinged vinegar together and take other (fresh) vinegar. In it, slake iron or steel 10 or 12 times. Now pour that to the tinged ⁂ ,put it into a large glass and distill all the vinegar *Per Alembicum* in the *Balneum*. What remains, will be the most beautiful redness of the world, because the tincture of the *Fish* stayed with the tincture of the *Star*. That us why the Fish is no longer of any use when the tincture of the Star is drawn off, because they both have the same nature.

With this tincture, you can work miracles, because it is indestructible just as heaven is incombustible; it changes everything into its nature and color, and thus it does to all stones & pearls.

THE PREPARATION OF THE SUN
OR, SALIS AMMONIACI

Ammoniac is of different kinds, that is, natural and artificial. Natural ammoniac is found in the earth, and it is again twofold, white and red. Both are extracted from hard clean ores. It is sweet of taste; its nature is hot and dry; and it is good for washing and purifying.

Further, there is also an artificial ✳ , and that is better for this work than the natural; it is also nobler than salt, and changes *Mercury* into water. When it has been prepared with it, grind it and put it in a humid spot to dissolve. With it, one accomplishes the solution

of ♂ and ♄ into living silver; and it is ☉ which the dryness of the fire has congealed. It is hot and humid and is a subtle spirit for the *Elixir*, for without it, it could not be done. How to sublimate it, I will yet teach you, although through it the *Corpora* are not transmuted white or red into another substance, except by means of other spirits, it nevertheless gives to the spirits their entrance and exit; it purges and cleans the *Corpora* of their blackness, leaving the spirits mingled with the bodies, while itself going away.

SAL AMMONICUM, OR THE PHILOSOPHERS' SUN

Ammoniac is the sun of the philosophers, which shines through all things and is the noblest Secret of the Hand, for by it I will teach you how to make the whole secret of the Tinctures, and in this way you can, if you wish, extract all other tinctures and use them for the great philosophers' stone. For the *Lapis Minor* you extract the things from the minerals, for the *Lapis Majori* from the herbs and *Species* that are not human; for like makes its like, a horse begets a horse, etc. Therefore, prepare like to its like; otherwise you follow wrong ways in this Art.

ADDITIONAL MANUAL WORK WITH ALUM,
OR THE *LANTERN* OF THE PHILOSOPHERS

Take 10 lbs of alum, and purify it by dissolving it.

Then take 2 lbs of white vitriol, 3 lbs of ⚲ , and 3 lbs of *Sal Alcali*, and these must be purified; also 2 lbs of *Sal Vitri*, 12 lbs of *Sal Gemmae*, taken from the apothecary; further, 4 lbs of the KEY and egg calx 4 lbs. All these have to be purified.

Now take the white of one hundred hardboiled and pulverized eggs, put into distilled vinegar in addition to 1 lb of ✳, also ⚲, *Alcali*, of each 1 lb; common salt, 4 lbs. Put everything into a jar, luted, and set into the *Balneum*, or in horse manure, for 6 weeks. Let it dissolve into water; then remove it and distill. First, the ✳ will come, then the ⁂ and the blessed white water that stays in the retort, preserve it till you need more of it, because it is also good for other works.

Now take the vinegar, the ✳, and the egg water, and imbibe with that your powder. Dry it in the *Alembicum*, so that it may stay pure. Do this till all the water has been imbibed into the powder. Set the water to calcine *In Tripode* for 14 days or longer, and regulate your fire in such a way that you can just keep your hand in it through the hole.

180

After this, take it out. Take 2 lbs of ✳, and
sublimate it through this powder, till it is fixed. Then
dissolve the powder 7 times in distilled vinegar, and
again congeal it in the *Alembic*, so that it may stay
pure. After it has been dissolved as often and has
finally been congealed, put it in the *Balneum* to
dissolve, or in a cellar, or in horse manure. Then the
Lantern is prepared, and you have an earthly treasure,
the whitest water one can see with one's eyes, and it
transforms all things into the White, no matter what one

cooks in it, be it metal or rock. With it one can make ☿
fixed in many different ways, and you, nevertheless, keep
almost all of your Lantern, and almost your previous
weight. Anything you wish to get pure and white, such as
pearls, put them into this water and you will see
wonders. It transforms all metals into Mercury, if they
are cooked in it in *Balneo*. I must not write more.

Now take the water of the HAND and 1 lb of the
impalpable powder of the KEY. Sublimate them, well mixed
together beforehand, 7, 8, or 10 times, each time with
fresh KEY. After this, pulverize it impalpably on a glass
slab with 1 lb of the Middle Finger, 2 lbs of the Fire, 2
lbs of ⊖. Sublimate 7 times, each time with fresh ⊖;
finally pound it and set it in *Tripode* to calcine for 30
days, heating gently the first 8 days, then gradually
stronger. The last 8 days, let it become so hot that you
would hardly hold your hand in *Tripode* for one *Ave Maria*.

181

Then take it out, powder it on a stone, then dissolve it into water in Balneo. Occasionally, it does not easily dissolve. That is due to the fact that it has not been sufficiently cooked during calcination, since one has to stir it once every day, what is at the bottom has to be turned uppermost; and if something had been sublimated up, put it down again and continue calcining. When it is dissolved, preserve it at once in a glass, well stoppered with something.

Now take gold. Purify it by dissolving and cementing, as you know how to do. Then dissolve the water of the Hand and keep it well stoppered in a glass. When you have all three of these waters, you have an earthly treasure. If you wish to conjoin them, take a glass *Recipient*, big according to the quantity of your waters, narrow above and wide below. Into this, put your three waters, seal as you know how to do; set it into the *Balneum*, and let it rise and descend till you see that no more spirits rise except a watery dew. Now take it out, powder it inpalpably, put it into a round glass with a long neck, put it sideways on ash, in a stove. Then you will see an evident miracle, that is, all the colors God has created in the world, before the perfect *White* comes; and then the redness from an increase in the fire. At first, one has to heat mildly, and gradually stronger. But when the white color appears, you must increase your fire somewhat till the glass becomes glowing. Should something rise, turn the glass over, and continue with the fire till you see the perfect redness which surpasses all the redness in the world. Thank God!

I have done it, but I never reached the end of the *Perfection*. But with it, you can well make *Projection* into all metals, because you have the *Qu. Ess.* of the ♁ of the philosophers, and you have subtilized, dissolved, and spiritualized the ☉. Then its color is increased a thousand times. Also, *Mercury*, has in himself the red color or *Tincture* in such an abundance, that it cannot be described. Therefore you must <u>yourself</u> experiment as to what wonderful *Projection* it makes. I am telling you in true charity that I and my journeymen never reached the goal of projection. That is why you must throw it on ♃ and no other metal. You may also dissolve your *Medicine* into a *Massa*, and carry it about with you, so that you can take the powder out of the glass, stir it into wax; and you can throw it upon whatever you wish, and close your mouth, by damnation of your soul. Amen.

PREPARATION OF COMMON SALT, OR THE KEY

Take sea-salt, pulverize it in a mortar, dissolve it in distilled vinegar made from white wine, filter and congeal it until a small skin forms on top. Remove this and keep it, because it is the *Spiritus* of the Key. Dissolve and congeal it again until the skins forms on top. Remove this again and the *Spiritus* is thus separated from the *Corpus*.

To 10 lbs. of this *Spiritus* add 1 lb. of ✳; dissolve in distilled rain water, and coagulate it again, *Per Alembicum*, with a recipient, until it is coagulated. Then take it out and pound it on a stone. Put your powder into a glass; dissolve and coagulate it again till it is fixed and no longer rises. With this Spiritus one does wonders and miraculous things in metals, especially with Mercury, ♃ and ♄. But this *Spiritus* must first be prepared, as I will yet teach.

ADDITIONAL MANUAL WORK WITH THE SALT
OR KEY OF THE HAND

Here I wish to disclose to you all the secrets of the salt. It is the greatest secret of all Arts of *Alchymie*, in which occur most of the errors made by people who are engaged in the Art, since most of the *Secreta* concern salts. Therefore, understand well what I am going to teach, for as you separate the *Spiritus* from the body of the common salt, thus is also separated the *Spiritus* from the *Corpora* of all other things. Yet they are not all prepared in the same way but each according to its work, and according to how they are needed and how one wishes to work with them. There are many kinds of preparation of the *Salts*, for in the power of the salts and their *Praeparatio* lies the whole Art of *Alchymie*. One cannot work with any salt unless the *Spiritus* has first been separated from the *Corpus*. After this, the *Spiritus* must be prepared according to the work in which it is needed. In addition, the *Corpus* must also be prepared in other works, because the *Spiritus* and the *Corpus* each have its particular power and cause contrary effects, as they are in fact *Contrair*, and, when they are separated, effect contrary *Operationes*, each in its own area. But those who try to operate with non-separated salts will work in vain, in spite of all their labors. They cheat themselves and all those who believe them. Neither will they achieve any kind of *Perfection*, for the whole Art consists in the separation of the salt and the preparation of *Sulphur*. There are many kinds of preparation, the same with *Sulphur*, according to what works are being done. The simpletons, who read in our books about *Sulphur*, believe

that we are speaking of the ⚵ which is dug up out of the earth; but we all have another *Sulphur* in mind.

There do not exist so many kinds of salts that there should not be as many kinds of *Sulphura* and their preparation. The *Sulphur* must be made artificially, just as the ✳ is made artificially. Now we will again return to our matter. First I taught you how to separate the *Spiritus* from the *Corpus*. Now I shall instruct you how to prepare the *Spiritus*, and to make it fixed, and also what you should do with it.

After you have removed the little skin, which is the *Spiritus* of the salt, and you have gathered as many of them as possible, dissolve them again in distilled rain water, the same in which you had dissolved your salt. Then congeal it again in a glass, or a glass vessel, over a gentle fire. Pulverize it in a stone mortar; add powdered ✳, mix them well. Put it into a glass with a long neck, lute well, put it to dissolve in horse manure or in the *Balneum*. The glass must be well closed above, or put a helm on with a recipient, everything well luted. Should any moisture come over, it will coagulate more easily if there is also an *Alembicum* attached to it than if it were only badly stoppered. This solution and coagulation *In Balneo* must be repeated 10 or 12 times, or more, till it is fixed. Each time it has to be coagulated over gentle heat. When it is fixed, the salt is clear and transparent, like crystal and hard. If it is put on fire or on a glowing *Lamel*, it does not smoke or melt away, and penetrate like lard through leather; and that is its proof that it is fixed. Now the *Spiritus Salis* is prepared, and it has also made the ✳ fixed together with itself; and they married and will never again be separated, but stay together and effect wonderful things together.

Put 1 lb of this salt into a Crucible and 4 lbs of purified ☿ with vinegar. Add also the salt below, but most of it above; set in for 6 hours in the fire, and it will coagulate into true ☽; your salt will neither be decreased nor worsened. It does not burn, neither does it fly away. Now put your ☽ upon the Cupel, as above.

If now you wish to bring ♈ to *Perfection*, purify it first of its blackness. Use *Amalg.* or calcination, and purify each time with vinegar and salt, till there is no more blackness in it. After that, drive the ☿ from it, *Per Descendum*, into another vessel. Now take the powder *Jovis* and *Sal Fixum*, put them into a crucible, as before, and let them drive for 6 hours; or pour the salt into it beforehand and let it drive. Afterwards add the powder *Jovis*, or both together. It is the same; I tried both. When it cools down, you find *Jovem* amid the salt. He is noble and a King. Sweeten the salt thereof and preserve it, because it is just as good.

Now finish your ♈ on the test furnace (or: cupel) with ♄ and drive it off, and you will find good ☽ in all tests. Aside from this, know that the *Corpus* of the common salt of which the *Spiritus* has been separated, is fixed in itself. It is prepared in the following way with *Sal Alcali*, *Oleo Tartari*, *Alumen de Roche* and *Marcasita*, that is:

Of this salt, take 4 lbs; of *Sal Alcali*, *Alumen de Roche* ana, 1 lb; *Marcasita*, 1 lb. Pulverize them fine, then pound them together on a stone with *Oleum Tartari*,

187

and make of them a thick paste. Put it into a glass
vessel or a Syburg jar; lute it tightly and put it under
the cookstove, where you keep a fire going every day. Let
it dry; when it is dry, remove it and pound it once more
on a stone with *Oleum Tartari*. Put it back into the jar,
close it tightly, and put it again in the fireplace to
dry, as before. Repeat this till the powder has imbibed
the *Oleum Tartari* and you have a dry *Massa*. Powder this
Massa to a fine dust in a mortar, put it into a Syburg
jar, lute it tightly all around, 3 or 4 thumbs' breadth.
Put it 3 or 4 times into a potter's stove; break the jar
open, and you will find it white.

Take it out and turn it into fine powder in a mortar;
let it dissolve in the cellar on a marble slate with a
glass beneath it into which it can drip. With this water,
one can do many wonderful things in the Art. With it one
can dissolve ☿ into a beautiful transparent water; when
☿ is dissolved with it or in it, one can dissolve all 7
metals into pure water with it or in it. And this is the
right preparation of the common salt, and it is the Key
on the little finger, because the little finger is the
KEY TO THE HAND.

HOW TO TURN ALL METALS INTO WATER

Now we will teach you to turn all metals into water,
since all of them must previously have been water before

they can be brought to *Perfection*. Purify the ☿ of its blackness, and purify it well without vinegar. Afterwards, dissolve it again in the aforementioned water, and you can dissolve in it any metal you wish. When it is dissolved, take sweet, fresh cheese curds. Boil them and skim the thick *Materi* off; let it settle. Separate from it what stays pure. Pour some of it into that which you have dissolved, and it will *Precipitate* into a *Massa*. Separate the water from the *Massa*. After this you must cleanse it of its blackness with salt and vinegar, and continue to work according to what purpose you wish to have the metals, to the Red or to the White, as different works indicate. But all metals must be water and purified of their blackness if one is to bring them to *Perfection*, or the great *Elixir* must do it: It has the power and nothing else.

Now I will teach you how to add the *Spiritus* to the *Corpus* that you have prepared, that is, in what way you must *Conjoin* them. When now the *Spiritus* and the *Corpus* come together and are united after their preparation, one can do wonderful things with them, since they have then a hundred times more power than they had previously; for after the *Coniunctio* of the souls and the body there exists a *Glorified Corpus* and a great *Elixir*. With it one performs great miracles.

First one cleanses ☿ of its blackness. After this, one must also dissolve in this water, ☽, which comes dry from the test furnace. Now take *Marcasita Lunae*; turn it into an impalpable powder and pound it well with salt and vinegar, washing it till no more blackness comes off. Now dry the powder at the sun, and afterwards grind it with good *Aqua Vitae*. Take ✳ water. Then draw the water off *Per Alembicum*, and immediately dissolve the powder again in the same water. Now take the water in which these three dissolved, and as much fixed *Spiritus* as the weight of the water in which the metals have been dissolved. Imbibe these spirits into this water, pound them, dry them, and imbibe them again; coagulate them again at the air, for it is the cold air that must coagulate them and the hot air must dissolve them.

Listen well to these discourses, how they are meant, it would take too long to elaborate further. Neither is it necessary for this work. When all this water has been imbibed, you have a precious Stone with great power to act upon ☿ and ♂. It can also be prepared for the Red, but that is not necessary; for when they prepare the aforementioned finger, they are making a great elixir. Should one work it to the Red, it would increase in power a thousandfold through the preparation. And this is one of the secret signs, and it is the little finger of the Hand, which is the Key of the Hand.

THE NATURE AND POWER OF MERCURY, THE FISH OF

THE HAND, AND ☿ THE FIRE

This is man and woman, this is sister and brother, this is earth and seed, this is origin and beginning of all metals; and without these two no work can be accomplished, for without seed no fruit can be born, and without soil no seed can grow. Accordingly, there has to be man and woman, water and earth, fire and seed, if any fruit is to be obtained. Thus these two are the origin of all metals, and without them no work can be Brought to Completion.

Now I will further instruct my child and describe the nature of ☿ and his infirmities, and in what way he is to be made healthy. Therefore, understand well what I say, for by means of this work I will teach you all other *Opera*, for all metals.

Mercury is a Spirit and also a *Corpus*, but the spirit is evident and the *Corpus* hidden and intangible. This is due to the *Spiritus* with which the body is covered, for the spirit is more powerful than the body. That is why the *Spiritus* takes the *Corpus* along when it gets into the fire. Mercury is cold and humid, cold in

191

the manifested and cold in his humidity; but in his innermost and in his hidden (nature) he is red, which is hot and dry. That is why the old masters called him an egg, because an egg is white, cold and humid outside, and within it is hot and dry; and when an egg is put in the fire, it will evaporate and burst. That is due to its humidity and cold. This is the reason why they call him the philosophers' egg, which people do not understand. Because of the coldness that he has in himself, he avoids fire; for two unlike things of a contrary nature cannot stand each other; one or the other must go. Because his *Spiritus* is stronger than his *Corpus*, the *Spiritus* leads the *Corpus* away from the fire. Now I immediately also show the virtues of Mercury.

Mercury is a slimy water in the earth, a subtle substance, an earth spirit. He is the same subtlety which the earth has in itself. That is why it is called *Arsenicum* of the earth. Therefore it has the power to produce fruit without the help of the earth; that is, he can perform works without any other thing, because the spirit of the earth and the spirit of a thing have the power to do a great deal without the help of the *Corpus*. But the *Corpus* has no power to do anything without the help of the spirit. By this understand well what I have in mind concerning *Mercury*, though *Mercury* may well be changed into ☉ or ☽ without the help of the metals. That is why the old folks called him the Serpent that begets itself and gives birth without the help of some

192

metals. Yet people do not understand this saying about the snake, because they do not know *Mercury*.

I have said that *Mercury* is a slimy water, of a white color, soft, natural, joined from within hot dryness and from without with cold humidity, more than with inner hot dryness; that is why he does not like fire!

Mercury is the *Ferment*, the yeast, or the sourdough, for yeast causes the bread or beverage to rise and to go over, and it gives them their taste. Therefore *Mercury* must be in all works like yeast (leaven), for without *Mercury* no work can be accomplished. *Mercury* is the beginning, the middle, and the end. He is the *Copulator*, the priest who brings all things together and marries them; because no fruit can come from like things; but unlike things, of two unlike natures, must be united. It is, however, not easy to unite unlike things. Thus, *Mercury* must gather all unlike things and marry them, if they are to bring forth fruit; for *Mercury* is in them as the universal Spirit, for he is the humidity of all things, also of the earth; just as blood is the *Spiritus*, humidity and life of men, *Mercury* is the life of the earth.

Therefore, all things that have got their nature from the earth are subject to him; also all that which the earth brings forth or has brought forth. That is why it is said that Mercury is the *Spiritus* of all things. Because of this the old philosophers say that in Mercury there are four kinds of saline juices, for all salts are made of four things, each contrary to the other. That is why all salts are poisonous, but one more so than another.

The old masters called *Mercury* "VIRGIN'S MILK", because *Mercury* is nourishment, food, and the dwelling-place of all metals, since he enters and, passes through all metals, just as mother's milk passes through all members of the child and nourishes them. *Mercury* is the *Nutriment* and mother of all metals; *Mercury* makes all hot *Corpora* cold, and all dry bodies moist; he also makes all cold bodies hot, and all humid *Corpora* dry. However, *Mercury* has to be transmuted into another degree, according to what the work is, that one wishes to accomplish with him. For *Mercury* has a wonderful nature. Wherever he is added or used, he is the best or the worst, depending on how the company is.

Thus one may use *Mercury* for whatever work one wishes to perform. If he is congealed, he continues to congeal; if he is dissolved, he dissolves further; if he is fixed, he fixes further; if he is multiplied, he likewise continues to multiply; if he is made poisonous, he poisons everything.

Thus have I taught and shown you what derives from *Mercury* and what he is. I have informed you of part of his infirmities as well as of part of his wonderful nature. Now I will also teach you part of his virtue and power and of the works he can do, from degree to degree, to his utmost, according to the understanding which

Almighty God has granted to me through his bottomless charity. Although one could not write all his degrees and powers in a book as big as the Bible, I will now here teach my knowledge obtained through the charity of God Almighty, and what I know, have learned and experienced in this matter.

First, what he can do when he comes raw from the mines. Subsequently, what he can accomplish when he is a great Sublimat; and then, what his capability is when he is congealed and dissolved, and, in addition, how to understand this. Finally, I shall make you acquainted with his sister and brother; and also how to make the ♃, by which ☿ must be brought into a state of health.

First I will say what *Mercury* can do inside and outside the earth, when he is still raw. I said that *Mercury* is the *Spiritus* of the earth; likewise he is the *Sperma* of the earth, and the seed of all metals. Of this seed, when planted in the earth, that is the ♁ - be it pure or impure, red or white - one of the seven metals is generated.

Consequently, *Mercury* is the beginning and the foundation of all seven metals. His intention is always to generate ☉ or ☽ but he is impeded by the lazy and impure earth or ♁, so that he cannot achieve his

195

purpose, while it is also true that the stench of the earth hinders him often to do so. For as a child in the womb is made impure by the *Materi* which it attracts — as the child may get a disease on account of the uncleanliness and *Corruption* of the place where it lies, which causes it to get such a disease; irrespective of the fact that the *Sperma* and seed were pure and clean, the child is nevertheless afflicted with and uncleanliness; it is due to the lazy mother and the *Corruption* befalling it - so it is with the metals which are likewise corrupted by the impure *Sulphur*.

After this, I will say and teach whether God has created all things of one or of two natures, and how, and what. God has put together and united all things of four contrary elements. He has put them together with their opposites; but these natures are perfectly united in all things, regardless if they are perfect or imperfect. Thus we find that either the manifest part in a thing is perfect, or the concealed part. But while they are cold it is the contrary. That is why it is easy to change the *Corpora* into their prime root in a short time, so that the concealed part can be made manifest, and the manifest part in nature can be hidden in the metals. For what iron lacks in its manifest form, is replaced by something in its hidden form during its transformation into ☉ or ☽.

Its hidden part is ☿, which replaces iron during its transformation.

196

If, therefore, ♂ is changed into ☿, its hidden part will become hot and dry, and its visible part, cold and moist. Bring its hidden part into manifestation, therefore, and make its visible part hidden. Then its hidden part will be cold and dry, and its manifest part, hot and moist; now it is perfect and will last in all eternity.

There were some who said that it was the hidden ♄ and ♃; others said ☽; and in this they spoke the truth. I am saying all this to show how all *Corpora* must be transformed; for in all bodies two manifest natures must be hidden, and two hidden natures must be made manifest. What is manifest, is imperfect and full of sickness; and what is hidden in the metals is fixed, good, and perfect by nature. That is why you must make its hidden part manifest, and its manifest hidden. Then you have a pure, perfect, and fixed *Corpus*, lasting in all eternity. It is a great secret to have the understanding of this, and to know the *Secreta* as also the infirmities of the metals, and in what way the metals came by them; apart from this, (it is also a great secret) to well know their natures, within and without, hot and dry, cold and humid, over and again back. It is the fire that I have in mind.

You can make them healthy again with natural medicine and with ♁, which I will teach you later how to make from green herbs, oils, gums, and water. Concerning this,

you must well understand my view, since in this chapter
you are being acquainted with all the infirmities of the
metals, and you will be taught to recognize all herbs, to
prepare all *Sulphura*, and to know all medicines. You will
learn what they require in their bodies, so that you will
learn how to make manifest one of the hidden elements,
and how to make its contrary hidden. Understand well that
I indicate all this, so that you should know nature. For
lead is cold and dry, its manifest part is quite soft,
and its hidden part is hot and humid.

In all things that God the Lord has created, the
hidden part is always contrary to the manifest part, in
softness as well as in hardness; for the soft is the
Contrarium of the hard, that is, when its nature is
reversed, its hidden part becomes manifest, and its
manifest becomes concealed; that is, the soft has to
become hard, and the hard, soft. This is meant for the

Corpora lead, ♃ and ♂. They are also composed of four
natures, since the manifest of lead is cold, humid, and
soft, while its hidden part is hot, dry, and hard.

Likewise with ♂. When they made him like lead, they

called him lead; but its hidden part is ♂ and the
former is manifest. And when you reverse the hidden

Saturn and make it manifest, it is ♂. Such is easy to

do with ♄, and then it is ☽, lasting in all eternity.

198

The body of iron is composed of four natures whose exterior is hot, dry, and hard, while the hidden part in its nature is cold, humid, and soft, like lead in its root. In no *Corpus* is there as great a hardness as in the manifest Mars, and in its hidden part there is an equally great softness. One becomes easily aware of its softness when it is prepared and reversed. After this operation it is ☿.

The hidden part of *Mercury* is ♂; its manifest part is a cold and dry water called *Mercury*. That is why his manifest part must be hidden and his hidden part made manifest. Thus he can be converted into fine ☉; or, if you wish, increase his color, and his coldness will appear and become ☽; or increase his heat and humidity, and it will become ☉ . Understand well what I say, what I intend to convey. Reflect on what I have taught you and how I have compared Mercury to an egg, which is white outside and red inside. Thus is Mercury in the center of his nature. Therefore, understand these teachings well. Enough of this.

Sol is hot and humid in its manifest part; dry, melancholy and like lead in its hidden part. Therefore, reverse all *Corpora* into this nature, for this nature is temperate. If you wish to cook *Mercury* till he becomes ☽, and till he becomes ☉, prepare your yellow-colored *Sulphura* that can tinct, and decrease his humidity till he becomes hot and dry. Then his humidity will be saturated and decreased, which means, to reverse his hidden part and root, which is ☉ and his color, which is solar, into *Sol*. Take my discourses to heart, listen and pay careful attention to them.

Venus is hot and dry, and is a brother of *Mars*, for his root is hot and humid like gold, and he is different from him in dryness of his *Minera*, which has become impure. Take of him his dryness, and prepare it with contrary *Sulphura* that are good and healthy for him. Then he will again get back into his nature and will turn into ☉ . Understand my discourses well, and pay careful attention to what I intend to teach you. It is of great importance and secrecy what I am telling you; it is also a great secret and deep wisdom. Therefore do take my discourses carefully to heart.

Mercury is cold and humid in his manifest part, driving and soft as water; and in his hidden part he is hot, dry, and hard without any doubt; for his manifest part is ☿ and his hidden part ♂ and the contrary. Now then, if we wish to reverse ☿ into his first water, the way to do it is first to turn him into ☽ through ♀ which helps him, that is, his brother and his sister. Understand my discourses: You must hide his humidity and reveal his dryness through ⟁. Then it will become good ☾ . Such is the first way.

Understand my teachings well, and if you wish to perfect the work, reverse ☾ so slowly till its manifest part becomes hidden and its hidden part, on the contrary, manifest. Understand well, these are two ways, active and passive. Then its hidden part will be hot and humid, that is ☉; and its manifest part, cold and dry, that is *Saturnus*. Do take care to understand my teachings well. I am addressing you in bad German, without veiled words. Know that ☾ is the root of ☉. Therefore, if you wish to change it into ☉, conceal its coldness. Then its heat will manifest. After this, cover up its dryness, and its humidity will manifest; and it will be good gold, everlasting.

Thus I have informed you of and made known to you the nature of metals and their infirmities; and how one is to reverse those and prepare them with each metal's ⟁; that is, according to what is the infirmity of a metal, prepare your *Sulphur*, as I will teach you.

Item, when ☿ comes out of the mines, it is living water, with which we must wash all our *Sulphura*, metals, and *Corpora*; and make all our *Corpora* humid. This means

201

that all their *Corpora* must be amalgamated and dissolved with *Mercury*, and *Mercury* must turn all hard *Corpora* into water or softness. Then one can pound and wash all metals with sea salt or distilled vinegar. This is to be done in a marble or stone mortar, and it has to be made so clean that no muddiness is left. When the *Corpora* are quite clean, drive the *Mercury* off *Per Distillationem*. Now you can prepare the powder with the ⚯, which I will hereafter teach you how to prepare.

Now I will further teach you how it is that ☿ becomes so poisonous during sublimation. The reason is that it is his nature to attract to himself all the poison he can get at. If he meets good metals, he wants to be the best; if he meets bad metals, the worst, etc.

This is the reason why ☿ attracts to himself all poison when he is sublimated through hot natures; for all hot herbs, salts, and all other things created by God that are hot to the 4th degree, are poisonous. I would like to inform you of the reasons, but this work does not require my doing so. Therefore, if he were to be sublimated through hot salts and hot *Species*, he would become poisonous; for all salts are poisonous, one more than another; therefore, the more he is sublimated through them, the more poisonous he becomes. One could sublimate so often through <u>Vitrioium Romanum</u>, or ⊕ *Album*, or alum, *Sal Gemmae*, *Saltpetre*, etc., that he would become so poisonous that, should he be put on a saddle upon which a man were riding, that man would immediately die as soon as his bottom got warm on it. But the salts must previously be dissolved nine times in distilled wine vinegar, and again congealed. And if *Mercury* were then sublimated it would die. It is possible therefore, to give to a man or an animal, poison the size of a bean, so that the whole body would become poisonous — all those who would eat of it would die. What should you now think of gold or ☽? And even if gold or ☽ were not medicinated but only *amalgamated praecipitation* in their raw state with ☿, and they were kept together in moderate fire, do you not think that the gold or ☽ would change ☿ into its nature? Yes, certainly, and that in a short time, within four weeks.

If then the ☉ and ☽ (even in their raw state) can bring about   the change of ☿ into their own nature, what will they do when they are *medicinated* with the noblest medicine of the world and are applied so abundantly? Such medicinated gold or ☽ is also dissolved in the water of the Hand, kept *in balneo* for seven days and nights in order to digest. Afterwards it is again congealed and then calcined *in tripode*, and this is done three times. What do you think of that? And even if they had not been medicinated but had by themselves been calcined and congealed, would it not have made a medicine all of itself? Indeed, yes. If you have intelligence, understand my words. If you do not understand, you are an

203

oaf. I have here taught you how to make the most precious of the most unprecious. Thank God Almighty that you have got this instruction.

<u>END OF THE WORK OF THE STONE OF URINE</u>

TRACTATUS JOHANNIS ISAACI DE URINA
How to extract all tinctures thereby:

You must know that all tinctures, white and red, or however they be, are all extracted in the same way, and that it is all a work toward the Red and the White. You must take a large earthenware pot or cask, well glazed, filled with old, clear urine. Set that on an *alembic*, which must be big, together with its receiver, and distill all that you can. Black feces will stay at the bottom. Calcine these for about three hours so that they glow moderately; then dissolve them in *aqua communi* and

boil it for two hours. After this, take it from the △ and let it settle; decant the clear liquid above the feces. Now, put it back on the fire and let it boil (the liquid) until a small flux appears on it. Then remove it

from the △ and put it in the cold air or in a cold cellar. A clear salt will crystallize. Remove this and

boil the ▽ down and let it sprout as before. Gather the salt again. Take the collected salts and dry them in an earthenware pan. Use earthenware because it can glow gently without melting. Now, distill it again in *aqua*

communi destillata and put it back on the △ for a quarter of an hour. Then take it from the fire, let the feces sink (settle), and pour off the clear as before.

You must always decant the clear while it is still warm; then boil it again until there is a small flux, like beans. Set it again in the cold air or in the cellar as before. Preserve the sal that has sprouted, and again boil the other ▽ or urine, which does not turn into sal, until there appears a flux, as before, until it turns into sal. Then dry the salt in an earthenware pan and preserve it until I tell you how to use it.

Then you must take all the processed urine that has been distilled with △, and if there is some impurity like an oil, or some yellow fattiness, skim it off with a spoon or with a feather, so that the urine becomes quite pure. Now, process this urine again with fire, into a glazed earthenware cask with an *alembic* and a *recipient*, and repeat this until there are no more *feces* in the earthenware vessel or cask. Always discard the remaining feces as they have no value.

After this, process (draw off) again in the *MB*, and some black feces will remain. Repeat this until the water goes over clear, without any *feces*, always throwing the *feces* away. Then take the aforementioned salt, which has been dried, put it into the clarified water and into an alembic. Close it tightly and set it on hot ashes for three to four days by which time the salt is dissolved into clear water with no *feces*, thus the tincture is now prepared.* This is now pure as before but devoid of its *fecibus* and of its coarse *humors*. It has now become so subtle, that it is inexpressible.   *(N.B. *Quando sal* *solvitur in aquam clarum absque fecibus tunc praeparatum est).*

Of this urine you should take six quarters (Viertheil) and three quarts of processed (drawn off) *acetum*, three quarters of *aqua vitae*, half a pound of common salt, half a pound of *salmiac*, half a pound of common *calx vitae*, mix all these together and let it dissolve into clear water without feces. Now you have a wonderful matter which turns all *calces corporum* into their first matter, that is, into ☿. With this water, one can draw out the blessed *Quinta Essentia* of ♂ and of all things which are red and white. Thus prepared, this water requires ten or twelve rides or trips (Reisen) and again drawn off, it retains all its power as if it had not been used at all. However, one must rectify this water.

HOW ONE CAN EXTRACT ANY TINCTURE THEY WISH WITH THIS WATER

Take *Sulphur* or *Auripigment* or Ochre or whatever you would like to extract a tincture from. Pulverize the matter finely and grind them together with distilled (drawn off) vinegar so it is like soap. Then, place this in a large vessel and set in the oven on ashes or sand and pour on it this clear urine, or the aforementioned water, so that the vessel will be half full. Then stopper the vessel with a cork and manually stir or shake it sufficiently so that the matters will mix (incorporate) well. Then, return it to the ashes or sand and, at first, give it a small fire thus heating the liquid. From time to time, remove the cork to allow air to enter or the vessel might crack; also, continually shake the vessel, by hand, to insure good mixing of the matters and to permit the vinegar (acetum) to penetrate well.

When you observe the vinegar to be colored well, decant the warm, clear liquid off taking care that no feces come over with it. Save this liquid for yourself, stoppering it tightly. On the feces, pour additional urine, as before, and some distilled vinegar, stopper it and shake it as previously done, to mix it. When a color (tincture) appears, decant it and add it to the liquid previously saved and sealed. On the feces, pour new Urine as often as will produce a tincture, that is, until no more color will appear. In this way, you will have drawn off all the tincture (or virtue) from the matter. Now you can throw away the feces or save it for use, if you know what it can be used for.

Then, take the aforegoing tincture in a recipient and evaporate off the moisture until a small flux appears on top. Then, let it grow cold and pour the matter, where the flux has appeared, into a pot that can be sealed. Lute a helm thereto and draw off all the moisture in ashes or sand, causing the tincture, either white or red depending on the matter used, to remain behind in the pot. This is the *Quintam Easentia* of the matter from

208

which you have made it. If you have added to it the white, so will you find the quintessence.

The White one will be as white as snow and the Red one will gleam like gold. In the foregoing manner, one can also extract the *Quintam Essentiam Mercurii* ⊥ ati in the Red or the White. Also from filings of ♂ or ♀ Nitr. out of Vermilion or out of AEre. usto. and also out of gold calx and silver, or out of ♀ in the quickest way out, of all things in the world.

NOTA: So that the Urine will be the strongest, you may want to throw in *Salmiac* and *Sal Commune Praeparatum* ana 1 "Quintlein" (a fifth part?) and then you will see the color as we have previously written about.

From these drawn—off or extracted tinctures, one can make *Cementa* and *cementiren* with it, which is a little known secret practice and art.

You can also make from this, Aquafort, that is as red as blood and as irridescent as a ruby. With this Water, man can do wonderous things, about which we are not permitted to speak.

THE FIRST WORK OF THE ANCIENTS

Know dear *Fulii*, that there were some ancients who took fine ☽, 3 Loth, (a Loth = ten grams) well processed

in the crucible, filed most subtly; and fine ☉,
cemented through *cementum regale*, also ground very fine —
one loth; well purified *Mercurius* — 8 loth. They
amalgamated all this well in an iron mortar with a steel
pestle, rubbing it thus for 12 or 14 hours. After this,
they put it into a stone or glass vessel cylindrical in
shape. (for example, a beaker - *hwn*) They set this in
sand with a heat that would not allow one to insert a
finger into the sand, and allowed it to stand thus so
that the humidity was drawn off. (evaporated) The
following morning, they found that the *materia* had become
hard. They then put it back into the mortar, and added
half a part of *Mercurius*, or 4 loth, so that the *intima*
was equal. They did this handiwork until the *materia* went
thus dry through a double linen cloth. Afterwards, they
put it yet another 8 days in the sand, in its vessel, and
ground the matter every day in the mortar for 8 hours
without stopping.

When the 8 days are over, take the *materia* and put it
in a small glass vessel. (a flat-bottomed Florence flask
was illustrated - *hwn*) Place a small piece of cut glass
on the mouth, and put a weight on it to hold it down. Set
it *in tripodem* and give it such heat that will allow you
to suffer your hand between the walls of the furnace and
the glass which contains the *materia*. The ancients kept
it thus for six weeks, day and night. At the end of the
six weeks they increased their fire somewhat - as much as
it is needed to keep lead molten - and they maintained
this heat until the saw the perfect blackness. Then they

rejoiced, for under the blackness the whiteness is hidden, and it is a sure sign that the *materia* has been well united in the beginning.

Note also, after the third day you must lift the cover of the furnace and see if some *Mercurius* has attached itself in drops to the glass near the top. If so, you must shake it down again, and if it will not fall down, remove the small glass and brush the drops down with a feather, so they will fall back on the *materia*. Then close the glass again, and do this each third or fourth day.

It is much better to give it little heat so that it does not ascend. It will then take longer, but that doesn't matter because (you will have) the surest for the least sure; for with strong heat, the work would no doubt spoil. Take you great care not to obtain the red color before the white; one color will change into another. There are also many strange colors, but pay no attention to them. Attention must be given only to the three colors that manifest in the work. First, the black; then the white and finally the red. Between these three colors, many other colors manifest, more than one can imagine! But pay no attention to these colors; they are flying spirits which are not fixed and which are poisonous. As long as you see the strange spirits, beware of the air, because it could kill you. The first red colors to manifest occur at a small regimen of the fire, or a fire increased a little.

In this Art, there is no other worry than to regulate the fire, but if you wish to go the surest way, keep your fire as small as possible. Then you cannot fail, however, it will take more time. Consequently, I advise and teach you, dear *Filii*, not to give too much heat, to prevent obtaining the red color before the white appears. If it <u>did</u> appear before the white it would seem to be as powdered bricks in the glass, like kernels of wheat or barley or a little larger, with mingled kernels of *Mercurius Vivus*. It would affect the glass. Then, all your work would be spoilt. This does not happen, however, except due to a too-strong fire. The correct color is not like brick but clear, dark and brownish—red. The color is more heavenly than red; and it appears with a good regimen of the fire, as will be taught hereafter. I am speaking of these colors, so that you should not go wrong because of ignorance, and not know what you are to do or not do.

PROCEEDING FURTHER WITH OUR WORK

If it happens that with a small fire something rises in the neck of the glass on the cover, open up and brush it down as I have mentioned. Keep it standing thus, day and night, until your materia has completely turned into powder. The powder should be grey and black, just like earth that has lost its moisture. And before you get to these colors, you will note many odd things, because the *materia* will become multicolored and piebald, with odd *macula* and spots, which all glowing and not glowing (or: burning and not burning). Toward the end of the last hour increase the fire so much that the matter glows nicely without melting. If you see it melt, remove the fire from the furnace as fast as you can, and let the matter cool down.

Now take it out and dissolve it in pure water. When it has settled, pour the pure off from its sediment while it is warm; for if you allowed it to grow cold, the Stone would crystallize of its own, and you would be unable to *clarify* it of its *fecibus*. You must do this — dissolve in water, pour off its sediment and allow it to sprout, and pour it off again, and boil it down, and let it sprout again - till everything is sprouted.

Then you must again dry the matter over a gentle fire, always stirring it with a little rod till it dusts, as mentioned before! Now put it back into the wide vessel of one thumbs' thickness and into a reverberating furnace, till your Stone no longer gives off any feces and stays clear and subtle, and melts on a hot tin like wax or butter. If it is taken off the fire, it must stand up and not dissolve even in cold and humid air. Then your Stone is subtle and fixed.

But if it should happen that the Stone should melt during *calcination*, it would not be spoiled because of it, but you would lose your weight; for if the Stone were standing long enough in flux in the fire, part of it would burn into glass, for the Stone is still alone and at that time does not yet have its *spiritus* with it, which could protect the *corpus* from the fire. Likewise, it does not have the soul with it, which would keep the *spiritus* and the *corpus* together in peace. When, however, spirit, body, and soul are united together, fire cannot turn them into glass, for it is then an *elixir* that surpasses all *elixirs*. Then it is a glorified corpus which is perfect. Then it is the indestructible *Qu. Ess.*, similar to the unconquerable heaven. And when you have thus prepared it, the Stone desires to receive the spirit and the soul. When you have got it to this stage, you have accomplished all that the ancients expounded covertly in their books by saying:

Take that which is closest to nature; from that draw our Stone, etc. And I have revealed to you all the things the *Philosophers* have kept secret. My child must know that this precious Stone is prepared in many different ways. This Stone can be used for any works one wishes, for it is now ready to receive whatever *spiritus* or soul one adds to it, either for the *Medicine* or for *Alchmy*. This Stone is not a chooser of persons, for it accepts everything one adds to it; for it is fixed and dry to the fourth degree, also cold, and all *spiritus* are volatile, hot and humid. That is why all spirits desire to be with this Stone; and that is the reason why some philosophers have called this Stone the Son of God, for He was no respecter of persons.

Item, the old sages have brought this Stone back to its first nature and utmost perfection. As they say: Four things arise from one thing. That is to say, the old people sought one thing and one root out of which four things originate. And when they were able to convert them back into one thing, the *Qu. Ess.* was achieved, valid in all eternity.

But if it is not, you must dissolve it again in the *AF.* and set it again into the *Balneum Mariae*, seven days. After this, again congeal it, and take it out again. And let it stand once again *in tripode* for eight days, as before. Then take it out, test it as before, and the more you dissolve and congeal it, the greater will be its projection. If the Stone does one to a hundred, and you dissolve and congeal it again, it will make a ten times higher *projection*. But I advise you to do it but three times, because the Stone would reach such great power and subtlety that it could not be kept in any kind of vessel. That is how *penetrating* it is said to be.

Consequently, I advise you to dissolve and congeal and calcinate it in tripode only three times. Then the Stone will become subtle and strong of its own, so much so that it is unbelievable. If ☉ is put into some oil, no one can express the abundance of its color. Yes, then it is of such great potency that if a man were to put three drops of *oleum solis* into a little rectified *aqua vitae*, he would retain his youth to the last days of his life as it is ordained for him. But this kind of oil must be made quite differently, not like the ordinary oil, which is made with *aquafort*. The *oleum solis*, however,

which is prepared as a medicine for the human body, is
made of two elements which you must draw from our Stone,
that is, the elements air and fire. With these you must
prepare your ⊙ oil.

Take ⊙ beaten thinly like gold leaves between
paper, and rub it on a stone with distilled wine vinegar
or with some water of ✳ or of the element which you
have drawn from our
Stone. When it is powdered finely, put it into a glass
pot. Into the same pot put the element which you have
drawn out of our Stone. Cover the pot and set it on sand
for three or four days. After that, open. it, and you
will find your gold transformed into an oil. Distill the
element from it, and *in fundo* you will find a golden oil.
That is the greatest medicine one can find in the world.

ANOTHER METHOD WHICH IS EASIER

Take our Stone in its coarseness (or: in its raw
state), such as it comes out of the *minera* of man.
Understand well what I am saying! Put it into a wide,
glass vessel and add the powdered gold leaves. Pour on
this some of our Stone, which must be old and well
settled and purified. Pour of the Stone, two fingers'
width over the ⊙. Set the vessel with ⊙ and the Stone
of summer into the heat of the sun. A white-golden skin
or oil will form on top. Remove it carefully with a
feather, in such a way that you move the matter as little
as possible. Put it into a glass. Proceed in this way

several times a day, removing the oil till no more oil forms on top. Thus you can obtain *oleum solis* with our Stone in its coarseness, as it comes out of man's *minera*.

Understand well what I have hinted at here, because there has never been a greater secret in nature concerning our Stone, which also, in spite of its coarseness, transforms ☉ into oil. And very many artists have sought this secret but have not found it. Therefore, be grateful to God, etc. If then this our Stone accomplishes this in its crude stage, just imagine what it will do when it is perfected and united with the spirit and soul, and is fixed subtle and fusible. Do ponder over my words, so that you do not do useless work.

Now we will again resolve to prepare our Stone for such great strength that it is unbelievable. You must therefore continue to dissolve the Stone in the water of the Hand, as I have taught you above. Dissolve and coagulate, and then calcine it *in tripode* as before. Do this three times and no more. Otherwise it would become all too *penetrating* and strong, as described above.

If, however, you wish to have your Stone *augmented* and *multiplied*, say one to a thousand, take 10 lbs of fine ☉ or ☽. After having prepared your Stone as before, melt it on fire in a crucible. After this, throw 1 lb of your Stone on it, and let them flow together strongly for a good half hour. Now pour it into a pewter vessel, or let it cool down of its own. Your ☉ or

☽will be brittle or crumbly, since too much medicine has been put on it. For this is the test of the elixirs: If one wishes to make *projection* with a certain thing, and one does not really know the projection, one throws the *medicine* on any metal one chooses. As long as the metal stays brittle, the medicine will accomplish yet higher projections. Now throw the brittle metal upon other metals, till the metals stay supple, etc. Keep this, for this is the end of the projection.

I said, you should throw 1 lb of the Stone on 10 lbs of ☉ or ☽, according to what kind of soul your Stone has, since your Stone is supposed to operate on unclean metals; 1 lb to 1000 lbs for good gold. And I instructed you to throw it on 10 lbs of ☉ or ☽. But this is done so that the 10 lbs of ☉ or☽ should also turn into *medicine*, because ☉ or ☽ does not require any medicine; but they will turn into a medicine which is better than your Stone. The reason is that while the Stone which you have thrown upon them is in itself the soul of the gold or the ☽, the gold or ☽ is medicinated. It is now a medicine and brittle like your Stone.

Pulverize the gold or ☽ there is, and dissolve it in the water of the Hand which is used for the White or the

Red, etc. Set it to putrefy into the *balneum* for seven days. Then *abstract* the water *per alembicum*, and congeal the matter. Remove it, pulverize it intangibly on a stone, and put it into a wide vessel of one thumbs' thickness. Keep it *in tripode* for eight days with a moderate fire, somewhat hotter than that into which you had put your Stone when you calcined it *in tripode* for this matter must be calcined somewhat hotter. Do this, dissolving, congealing, and calcining in tripode, three times, and your matter will be stronger and better than your Stone. The reason is that your Stone is the soul of your matter, and the gold is the *corpus* of the souls. When the gold has been turned into oil, it has a hundred times more power, as said above.

Gold, however, is not just oil but also a medicine, just as good as is the Stone. Example: Give poison to someone, as big as a bean, an evil, strong poison. That man will die immediately, because the poison courses to the heart and through all arteries, including all flesh and the whole abdomen. It poisons the whole body. And if a man were to eat of an animal to which poison had been given, as I have here described, all those who had eaten of it.

I know of no thing in the whole world which would be as good and wholesome to our nature as this divine Stone of the philosophers.

Now we will again give information on how we are to prepare our Stone, which is at first a dead *corpus* but has been glorified and made pure, and suitable to set in it the spirit of life and the perfect soul and to make them eternal.

If you wish to make this Stone come alive, you may bring it to any body you wish. You can make of it a *Lapis Philosophorum* or *Qu. Ess.*, which cure all sicknesses, which sustain man's body in full health and let him last without decrease of the body till the last termin of everybody's life, as we heard above. But if you wish to make of it a medicine for unclean metals, you must take it to metals, for a horse makes a horse, etc.

Further, then, in order to achieve our purpose, my child should take ✳ *dissolved in AF.* , sublimate it 4 or 5 times through vitriol and salt. The more it is sublimated, the greater will its *projection* be. Following this, ☿ thus sublimated is to be rubbed to a powder on a stone, and this is to be put into a wide vessel, one thumb thick. Set it *in tripode* to calcinate for eight days, however only with a gentle coal fire, so that you can keep your hand over the fire for the length of an Ave Maria.

Now take it out and dissolve it in *Aquafort* made of saltpetre 1 part, *vitriol Romani* 2 parts, cinnabar ½ part, ✳ ¼ part. From this make strong water (aquafort), as you know, and rectify it as it should, etc. Then dissolve as many ☌ ☉ as you have pounds of

☿; dissolve each in a separate glass, and when☿ and ☉

are dissolved, pour the two waters together and set them

in the *Balneum*. Let them stand for seven days to unite,

that is, the *spiritus* with the *anima*.

After this, rub your Stone intangibly on a marble and

add the powder to the glass which is standing in the

Balneo with ☿ and ☉. Let your Stone dissolve with the

spiritus and the soul. Then all three will turn into

water; let them stand dissolved into water for three or

four days, so that they may become well united and marry

each other; and give more *spiritus* than you have *corporis*

or stones, because the *corpus* will not absorb more

spiritus that it has a right to. Now distill the water

per alembicum out of the *balneum*. Pulverize them on a

stone, put them into a wide vessel, one thumb thick, set

the vessel *in tripode* to *digest* or to *calcinate* for eight

days and nights, with a moderate fire. Then take it out,

put the matter into a glass pot, *lute* a small glass on

the mouth of the pot, and set it to *sublimate*, since I

have taught you before that you should take much more

spiritus than you have *corporis*, and in this *sublimation*

the *corpus* will let the *spiritus* of which it has too

much, go.

Let it stand in the pot for three days and nights

with a good fire, as is necessary for *sublimation*. Then

221

take the pot down and take the Stone out. Test it on a copper, glowing sheet; see if it melts like wax, spreading on the sheet, penetrating into it like fat into dry leather. See if after the sheet has cooled down the spot where the Stone has spread is good gold in all tests (assays) — then your Stone is valuable and all ready and accomplished.

If then someone has married one of these daughters, he will never again be in want, but it has to be achieved by dint of great effort and care; and it would really be necessary for a good and experienced alchemist to have all these daughters in marriage, to know well and understand all these operationes, to enable him to distinguish between good and bad; but enough of that.

Now to revert to our thema, that is, to our Stone of the free art. Open your ears, then, and listen; open your eyes and look; open your understanding and take note, for I will reveal to you secret matters which no one has as yet revealed. I will disclose more to you than I have been told to. If you have the least bit of intelligence you will understand it, as otherwise God will not give it to you.

Listen: Before our Stone has become sufficient (or: adequate), it is already alive; and when it is found, it is dead; and everyone sees it and holds his nose before the *Materi*.

THE STONE OF URINE

A GOOD AND SINCERE WORK OF:
ISAAC HOLLANDUS

Before our Stone becomes (or: comes into existance), it is alive; when it is found, it is dead; everyone sees it and holds his nose before it; it lied on top of the casks or vessels in which it is kept for a long time, and one and all hold their noses before the *Materi* or stinking air from which our Stone is drawn. The poor have it as well as the rich, little children as well as older people. It is a child's play and a woman's work; and ignorant men have looked for it in excrements and have not found it. For when you are alive, the Stone lives with you. That is the reason why one cannot draw our Stone out of excrements, since our Stone possesses the four elements perfectly; yes, it is more wonderful than anything on earth. For man is the very best, which God has created in this world in his likeness (or: in his image). If you have some intelligence, listen.

Our Stone has a strong smell and bitter taste, like urine, and it is found everywhere in superfluous quantity. All animals also have it, though not as perfectly as man. Without our Stone nothing in this world can live. I am telling you enough, if you will only understand; and if you do not understand, God Almighty will not grant it to you; and even if you do not find it, it is nevertheless found.

Our Stone is in all things that grow out of the earth, and it is also in the earth, likewise in ditches and also above the earth. Should God then provide that you may find it and know its nature, we will inform you how to extract it and how to proceed in order to draw the Stone of it, of what color it is, what it must look like when it has been made; likewise how to handle it to prepare it.

Item, our Stone costs little and can easily be found everywhere, in all street corners, in all secret chambers; on dung heaps and in caverns and vaults or in stables, there is an abundance of it. It grows and greens in all places where its water is found and where it lies quietly. Our Stone also grows out of the foul, stinking *Materi* in which it is white and clear, just as glass grows out of the foul earth and is also beautiful and clear. Therefore the ancients and wise men write: Our Stone purified itself and separates itself from all uncleanliness. The ignorant ones, who do not understand this, rebuke the ancients for having said this, and

believe that it is ☿. And further, our Stone rises above all *Feces* and ascends up high, where it collects.

Item, if you know the Stone, take it in its coarseness, congeal it till it is thick, and guard it from all metals, because the Stone would turn into weeds, for it is their nature to make all things pure and clean. When you have made it thick or have congealed it, you may draw from it the two elements air and fire. The third

224

element, earth, lies burnt black, like coal, *in fundo* of the vessel. In the black coal there is hidden the Stone of the old and wise philosophers as also of the sworn Masters. Pulverize this black earth intangibly. Put it into a wide vessel *in tripode* to calcinate for four days, glowing in moderate heat, so that the matter stands between glowing and not glowing. But the last day, let it glow nicely, but not too much, as the matter must not melt; for as long as our Stone is not pure, it is combustible, and the Stone together with the foul *Materi* would burn to glass if the *Materi* were to reach the melting stage. That is why the ancients forbid heating any matter too much till it is pure and clean and united with the *Spirit* and the soul; for the *Spiritus* preserves the *Corpus*, so that the fire cannot either burn it or harm it; and the pure, clear corpus protects the spiritus, so that it stays in the fire and does not fly away while the body is fixed; and thus it does not let it fly away from it. The spiritus is incombustible. That is why the spiritus does not allow the corpus to burn, for they are one by means of the spiritus and the soul on the spiritus and with the body. For although a pure spiritus and corpus are joined together, the fire would nevertheless separate body and spirit, and the spirit would escape. But when the soul is joined to the body together with the spirit, and they are all pure, they are one. Then neither fire, or water nor anything in the world can destroy them, for it is a perfect thing.

Item, when the earth is thus calcinated, take it out of *tripode* and dissolve it in common, distilled water;

let the feces drop, and as long as it is still warm, pour the water above off into a wooden or stone vessel. Do be on guard against the metals, otherwise the blessed Stone will be spoiled and *corrupted*. Now the blessed Stone will sprout beautifully and purely, and grow like grass out of the earth, ever more and more.

Now pour the water of the sprouted matter into a stone vessel which must not be coated with lead, or into a glass vessel. Boil the water, and again pour it into a wooden or stone vessel. Let it sprout again, and each time something has sprouted, boil the water down till everything has sprouted.

When everything has been boiled down and has sprouted, dry it over a gentle fire, stirring with a fine rod till the matter is so dry that it dusts. Now put the matter into a wide vessel, of one thumb's thickness, and set it *in tripode* or a reverberating furnace. This is the best and last calcination, which is to last three hours; the first hour with a gentle fire, the second with a stronger fire, and the third hour the fire must be heated so strongly that it reaches the stage between twenty and thirty times through it, he has the power to accomplish such feats, and he would stand the test.

It is not necessary, however, to bring *Mercury* thus to *Perfection*, since it is dangerous to do it on account of the poison; for if the pot got cracked, it would be dangerous for those who do not know him; since one must heat so strongly that the salts come in flux before ☿ will evaporate from them. Afterwards he does not completely rise but stays below in the vessel to cook with the salts. Yet to do this one has to sublimate him forty times. I have done it myself with my own hands, and afterwards drove off 20 lbs on a test, and I did not lose one pound of it.

I do not, however, advise you to do this because of the worries one has with it, since the last times one has to give such strong heat as I indicated. Also, there are many ways that are easier to bring *Mercury* to *Perfection* with ♁; I only wish to show you his powers, for when he is thus sublimated, one can perform miraculous things with him. It would be a pity if some people knew about it and understood it, as they would then perform miracles with him. That is why I cannot write to you about all his powers which I know and which I have tried, but I will relate part of what he can do when he is sublimated.

It is known that in the Art of *Alchymia* there is much fraud, and many imitations of ☉ and ☽ are made which are supposed to stand three or four tests; and yet they are finally false. If you should have any doubt concerning this, take some powder of the sublimated Mercury; put some of the ☉ or ☽ about which you have your doubts into a crucible, and melt it; throw your powder on it, and it will immediately reveal its nature;

if it is false, it will reveal itself and get back to its first nature as soon as it melts. But if it is good, it will stay good. Throw the same powder on a stone; if it is false, it will immediately break into a hundred pieces, like salt.

Dissolve (melt) ♃ and ☽ together ana; put it in a test with *Saturnus* and throw this powder on it. It will drive (or: flow) as if there were no ♃ in it, and it will be of great benefit, as I have tested myself. If you have ☽ that is not malleable, throw this powder on it, and it will become malleable like fine silver; throw it on ♂ or steel, and it will become malleable and soft like lead; and one can test all things with it.

Now then, all *Sublimationes Merourii* are beneficial to his recovering his health, provided he is sublimated through hot, dry things. Likewise, all *Congelationes* are good for him, because all things that congeal Mercury, be it green herbs or salts, or ♃, or lead, or gums, or smoke, or sulphur, etc., are absolutely curative or healthy for him, and one should congeal him so long that he can stand in the fire.

There may well be more direct ways with sulphur; but when Mercury is congealed, he has powers and effects which he does not have when he is sublimated. It is not necessary, however, to relate them in connection with this work. When Mercury is dissolved, he does other wonderful works, and one does with him what one wishes; for when he is dissolved, he dissolves the seven metals. And, what is more, he dissolves all things grown out of the earth and everything the earth has given birth to, not everything in the same period of time but one in more time than another. Yet in time Mercury dissolves all things, also with the help of other *Species* that are added to him.

Thus Mercury is the Lord and *Spiritus* of all things, for he shuns nobody and nothing in the world created by God Almighty, except fire. But with the help of the precious *Sulphura*, which the sages have invented and made by the Grace of God, and after they have well considered all powers and degrees of Mercury, as well as his infirmities and the causes thereof, and what he is, the masters have prepared a medicine for him; that is, a precious *Sulphur* made of many hot herbs and other *Species* which they used for it; and they have cooked and fried him so long in it that he could get into the fire and stand it. I will teach you yet how to make it.

Take small pearls, dissolve them with Mercury — all of them have to turn into water. Now pour on them fresh cheese curde, as I have taught you in the Key. The *Materi* of the pearls will precipitate to the bottom. Wash it clean till all the *Aqua fort* is off, and you have a viscous oil. Of that you can make pearls as big or as small as you wish.

Have the form made of silver, gilt inside. Put it into the blood of a buck, mixed with nightshade water, in which they will become as hard as they were before; and they will get such a pure lustre that is more marvelous than that of all other pearls; for during the solution they become so clean of all spots, so purely oriental, that no uncleanliness whatsoever stays in them. The same happens to all other precious stones dissolved in this oil.

Item, in the second instance you may take crystal and dissolve it. Then take the tincture which I taught you to prepare from ⊕ Romanum, in the foremost finger of the Hand, upon which stands the star. Dissolve it in water together with the crystal, and conjoin them in the same way as you did with the pearls. Form them big or small according to your wishes, make them hard like pearls, polish them with OIL of BIRCHWOOD made *Per Descensum*. With that oil all manmade stones have to be polished, and whenever they are polished, the oil has to be spread on lead. After being polished, the stone is like a beautiful, precious ruby, looking as if it were worth one hundred crowns. Nobody can distinguish it from a ruby, unless it were subjected to the highest test; but there are not many people who know how to do that, yes, hardly one in a region.

Thus you can counterfeit all kinds of stones from crystal and give them the coloring you wish the stone to have, but the tinctures and colors must be dissolved with the crystal in the manner which I taught you concerning the ruby. All kinds of glass can be dissolved in this way, like crystal, red, blue, yellow, green; glass of all sorts. I have seen wonders made of it, wonderful things made of glass and crystal, which great Lords possessed and considered more highly than fine gold. They were formed with foreign animals and wonderful. They were considered precious stones and were set on feet (pedestals); also on exquisite bowls, on treasure chests; and nobody knew what they were, except those who know to make the supreme test of everything.

One can also take little pieces of rubies, or sand, or other stones, (no matter which), dissolve them in the aforementioned manner, form and harden them as said above. They will become as hard as before, but much purer and more beautiful, for nothing impure remains in them. And in this there is no fraud, because they can stand the highest test; and one does indeed find enough little pieces of precious stones, and therefore it is not necessary to accept or to make imperfect ones.

But this is enough now of rocks. One does indeed make many wonderful things from *Mercury* after it is dissolved, so that it would be a pity if many a man knew about it.

Item, all metals that one wishes to bring to *Perfection* must be dissolved in Mercury, otherwise it is sheer fraud. Then they must be cooked with *Sulphur*; but you have to prepare the *Sulphur* according to whether the metals are healthy. Understand me well! I will teach you the *Sulphur* with which you must make *Mercury* healthy, and in connection with this *Sulphur* I will teach you how to prepare all other *Sulphura*.

Mercury is cold and humid in his outward nature, and in his innermost he is hot and dry. Reverse him, therefore, and he will become healthy. Consequently, one should remove from him his cold humidity. This must be done with hot herbs or oils of which you must make the *Sulphur*.

Take alum de Roche, calcine it as is customary. Now take the peels of bitter oranges. Put them into a glass vessel, pour good distilled wine vinegar upon them, boil them till all the vinegar is boiled away and the powder is quite dry. Add this powder to the said powder of alum, together With three oak apples (gall nuts) that are pointed; pulverize them. Then take sloe herb, pound it quite fine, and force it through a cloth with good distilled vinegar; pound it again till all of it has gone through the cloth together with the vinegar. Now dry it in a bowl and let it dry at the sun. You will now have a fine powder which you should add to the other.

Now take cinquefoil and treat it like you did the che-lidonia. Add it to the other powder, and the more hot herbs you gather, the better you can get *Mercury* to die. Now take all these powders pounded finely together; put them on a grinding stone, imbibe them with pig's gall - that's the best - let it dry till you can pulverize it again.

Add to this dried human blood, the blood of roosters or hens; imbibe them together with animal gall as before, and let them dry together on a glass slab. Do this eight or nine times, the more the better; and finally let it become quite hard and dry, so that it can be turned into a fine powder.

Now take Roman, red-calcined ⊕, and as much cinnabar as the vitriol weighs. Pulverize them quite finely together, put them into a glass, pour good *Aqua Vitae* thereon; then distill the *Aqua Vitae* off Per *Alembicum*. Repeat the drawing off and pouring three or four times.

After this, take the poured off *Aqua Vitae* of these two matters. Imbibe the powder therein on a grinding stone to the consistency of a pulp. Let it dry again on the glass as above. Do this twelve or sixteen times, the more the better. The last time, let it dry hard, so that you can powder it in a mortar and pound (grind) it on a stone.

Now take *Mercurius Sublimatus*, or calcinated *Mercury*, 1 lb; of the prepared, pulverized *Sulphur*, ½ a pound. Mix them together and put them into a SYBURG jar, well stoppered. Put them where you heat every day, and let them get heated as much as you can, because *Mercury* will not fly away while he is lying with his brother and sister, arm in arm, mouth to mouth; and they will surely keep him there, so that he can stand the fire; for his brother is hot and his sister is dry. And they are the hot, dry *Species* of which the *Sulphur* is made. This is the *Sulphur* of the philosophers. This is the same of which the philosophers write, but the ignorant imagine that they mean sulphur. This *Sulphur* I have often made.

Item, one can also make *Sulphur* of hot, dry oils; or hot, dry gums, or hot, dry seeds of herbs, but you must always add calcined alum. Sometimes you must add hot and humid, sometimes cold and dry; occasionally cold and humid; at times dry; sometimes hot, sometimes cold, sometimes humid, according to the infirmities of the metals that you wish to reverse outside, inside. Sometimes only one nature must be reversed in a metal; in another instance, two. That is why the *Sulphur* must often be made in different ways, because there are many kinds of infirmities in the metals. You can also transmute the metals into another nature and you can do this with *Sulphur*.

One can also change ☉ into ☽ with *Sulphur* made for that purpose. Nevertheless, that is contrary to nature, since nature is always aiming at the best. Nevertheless, it is possible to do it, that is, to *Transfer* one thing into another nature by means of the *Sulphur* made through the Art. Therefore, whoever can transmute ☉ into the nature of ☽ can also translate *Luna* and all other metals into ☉ ; for it is much better to do that than to

234

change ☉ into ☽, because nature always desires what is better. This is also the reason why one must prepare the *Sulphur* according to the works which one Wishes to do with the metals or to *Effect* upon them. But whatever *Sulphur* you wish to make, be it hot or cold, dry or moist, you must always have calcined alum. It must be in the heat, coldness, dryness, and humidity of the Sulphura, for alum is the Lantern in the seven metals.

Let us now deal further with *Mercury*, who lies shut in the stove with the *Sulphura*. You must give fire day and night, that is to say, on the jar; and keep it always hot. Every fourteen days you must go over to it, take out two or four ounces, put it with ♄ on the cupel and let it drive, (extract). And lo! If it drives (moves about or floats) on the test, you can take it out, our clean water into a bowl, and pour *Mercury* with the *Sulphur* into it and stir well. Now Mercury will settle down at the bottom, then you must pour off the water with the *Sulphur* above, into another bowl, five or six times, till the water runs off pure and becomes sweet. When all the *Sulphur* has been washed off it, dry Mercury over the fire, and drive it off on the test, as on ☽, and you will find fine silver in all trials. Let the water containing your *Sulphur* evaporate on fire, and you have your *Sulphur* back; but it has been somewhat decreased because of the washing.

This is the first secret sign of the seven secrets. It is called the Fish,which is compared to *Mercury*, and it is the most secret sign among all seven secrets, and it is the beginning, middle, and end of all perfect works, as I taught you before. I am therefore asking you not to let this get into the hands of the ignorant and sinners.

SUBLIMATIO MERCURII WHEN ONE WANTS TO DISSOLVE IT

If you wish to sublimate *Mercury* in order to dissolve it, let salt melt in hot water, and let the water stand over it for three hours to dissolve it. Now pour the water off and congeal it. Through this, subiimate Mercury as often as you wish, each time with fresh *Cement*; and do not take what is not dissolved in three hours, as it is not good for this work. This salt is good, although you have not much ✳ with it. It will probably dissolve when it is calcined as it should. When this Mercury no longer shines, it is sufficiently sublimated and pure; and the test is when he no longer blackens, although he is putrefied.

Item this test: Put ☉ on a glass slab, and if everything together dissolves so that no feces remain, it may well be putrefied; but it must first be congealed before it is put on the slab and putrefied. If any *Feces* remain, put it back into *Putrefaction* and feed it, and see if it does not require more; and you will feel that when the little pieces stay inside whole and no longer dissolve. Then congeal your *Materi*, which is now good and ready to work with.

A WATER WITH WHICH TO CONGEAL MERCURIUS AND MAKE HIM PERFECT

Take *Vitriolum Romanum*, cinnabar, *Sulphur* and alum de Roche, white lead, litharge, *Magnesia*, *Marcasita Ana*, 1 lb, pound them on a stone to a subtle (or: fine) powder; take 1 lb ♃ and stir it to ashes, as *Mini* is made. Pound this powder with salt and *Aqua Vitae* and dry it again at the sun or over a small fire. Then mix this powder and pound it on a stone while moistening it; dry it again as above.

Now take *Sal Commune* and *Vitriol Ana*, powdered and well dried over a fire. Make an *AF.* of them. Pour this *AF.* on Mercury in a glass; put it in warm sand till Mercury congeals; after this, *Abstract* the *AF. Per Alembicum*, and you will get your *Aqua fort* back and better for the same type of work. Further, take ☿ coagulatum and melt them together; pour about 2 lbs other dissolved lead, and let them stand for one hour together over fire in order to dissolve.

Afterwards, have at hand a pot with a long neck and a fitting lid. Into it pour pitch, resin and 🜍 *Ana*, 1 lb.

Let them melt together; then pour the ☿ and the lead together into the pot, close it well with the lid so that
it does not burn toward the outside; afterwards let it
cool down; then drive the *Materi* off on a test, as one

Refines. You will find good ☽ in all assays, and you
will lose little of the weight of Mercury, except that
which is impure, as I myself have done more than a
hundred times.

ANOTHER COAGULATION OF MERCURIUS

R.- *Alumen Roche* 1 lb, *Vitrioli* ij lb, verdigris ʒ

iiij, iron oxide (hammer scale), tin ashes, ♁ litharge
ana 1 lb, common prepared salt, child's urine, ana 1 lb,
add to it finely chopped pig's hair. Of this distill an
AF., and put in the *Recipient* four pig's galls.

This *AF.* is to be put on ☿ , in a glazed pot, on warm sand or ashes, till Mercury is congealed. Put this congealed Mercury in a glass; put on it half a pound of lead, melt them together; then have at hand 2 lbs molten *Saturnus* in another crucible. Pour that to the lead and Mercury, and let them stand molten in the fire. Now take a pot with a long neck, or a Syburg jar, well luted, and a cork that closes its mouth tightly. Put in it pitch,

239

resin and $\overline{\overline{\textstyle\bigtriangleup}}$, human hair, horn, cow's claws cut up, pig's muck ana 1 lb, spikenard oil and linseed oil ana iiij. Melt everything together in that jar, and when it is melted, put it into a crucible. Then pour the molten *Saturnus* and *Mercury* in it, and stopper it immediately so that the flame should not come out. Let it stand in the heat for 24 hours. When it is cold, take it out, drive it off on the test, and you will have fine ☉ in all assays.

THE PHILOSOPHERS' POT

This is the pot of the philosophers or sages, of which they speak so discreetly in their books and parables, so that nobody can understand it except those who are familiar with and have sworn to the Philosophic Hand. That is why I advise all who wish to roast, boil, or cook the Egg of the sages, that they should take care lest the shell bursts or cracks in the fire; for if the Egg were to burst, all the poison described in the pot would get out, and would kill and destroy all persons near it. Nor could they be helped with medicine, for in it (the Egg) there is the most evil poison that can be found in the whole world. That is why I wish to advise all of you who are not familiar with this Art, nor have sworn by it, that you do not dare cook, boil, or roast the Egg, for you fare badly and would get all the troubles contained in this pot.

This is the fruit brought forth by the seed thrown into the earth; this is the fruit of the philosophers; this is the faithful Stone of *Alchemia*; this is the earthly treasure and an earthly God, in whose hands lies the spiritual and temporal laws. He holds the whole world in his hand and gives people to understand that they are likewise to possess all things in this world. Whoever wishes to cook, boil or fry this egg, let him most diligently examine the Hand of the Philosophers, soas to probe it with his intelligence; thereafter he may cook it

TWO JARS

In these two jars there is the distilled *Aqua Fort* of the philosophers. Into this water one should put the earth and throw the seed, then put it into a glass with a

long neck, close
Now put it in h
will grow.

Prepare
calcined ta

water must

matters;

this wat

which a

and po

of ai

water

come

clo

F

boil them together. Now pour you will
thus draw off its phlegma. T till it
gets hard. Now take ⊖ 2 lbs re warm
water, filter and congeal it take of
the long finger 1 lb. It must blimated
through salt. Mix it with the s, put them
together to sublimate. You mus with them
till it is fixed. After this, ine for
twenty-one days, as you know; solved, cle
it of its Fecibus. Then coagul ke it fusib
Then work with it.

TINCTURE FRO

Stir lead to ashes and mak of it; t lb
of it. Take 3 lbs of the KEY pre of
the FISH, 1 lb. Mix them to an or sublim
sublimate them together. Then th ble powde with
it the Q.E. or Tincture of Satur will b than
gold in all works. You can Separ ich is way as
was taught above in regard to the t in the Fish
is no longer of any use, except t ar. B . The Qu.
Ess. or Tincture is that which imme hrow gulates the
Fish to redness and causes it to be ia ed.

TINCTURE FROM ♀

If you wish to extract Qu. E neris (the
quintessence of Venus), calcine th the Key, with the

244

STAR, and with *Sulphur*, and prepare an *Aes Ustum* as follows:

Take pure urine of men, or good distilled vinegar, or old, pure, sour beer; put the powder in it and boil it till the *Liquor* is colored. Then pour it off and add other *Liquor*. Proceed as before till no more *Tincture* is extracted. Now put all the tincture together into a retort with an alembic, and distill the humidity off it *Per Balneum*. Then the most beautiful redness will remain, which is better than gold. Sublimate the long Finger four or five times through it, and dissolve it in the cellar on a marble slab. After this, coagulate it again *In Balneo Per Alembico*. With this you can now coagulate the soul of *Mercury* to the most beautiful redness of gold, and it can also easily be fixed in this way.

The other way is the following: Take 1 lb of the powder, 1 lb of the Key, mix and sublimate them. Proceed with this as has been taught concerning *Saturnus*, and you will have the *Qu. Ese. Veneris* better than gold.

ANOTHER WAY

R. ♀ 3 lbs, Fish 9 lbs. Prepare an *Amalgama*; dissolve it in water of the Hand, set it for six weeks into *Putrefaction In Balneo*. Then distill it in a boiling *Balneum Per Alembicum*; set it in sand, and distill yet more spirits of the (symbol missing). When they have gone over, increase your fire and the Fish will sublimate and the *Qu. Ess.* be contained in it. Take the sublimate out

245

and put the Fish with the *Tincture*, pounded to an impalpable powder, into the *Balneum* with good distilled vinegar. Now the *Qu. Ess. Veneris* will rise into the vinegar and the Fish will settle at the bottom.

Pour the colored vinegar off, add other vinegar, and proceed as before till your vinegar is no longer tinged. Now Abstract all the tinged vinegar *Per Alembicum*, and the most beautiful redness will remain, with which no gold can be compared. You can dissolve this tincture in water of the Hand; that is better to the Red. Congeal it again. You may do this so often, till you have achieved such great subtleness that it would well be worth a kingdom. Then you have the *Qu. Ess. Veneris*, of the Star, of the Fire — which is Sulphur — and of the Fish, all together.

STILL ANOTHER

Take copper filings, boil them in good distilled wine vinegar. To every pound of vinegar add 1. ℥ ⚹. Set it in *Balneum* for six weeks; then add the helm, and the ⚹ will sublimate with the ♀; a grit will stay at the bottom like a salt. Calcine it without any addition till the matter turns red. Now take it out and pour distilled vinegar over it. Extract the tincture as before, and you have the *Qu. Ess. Veneris* by itself. With it you can

Cement whatever you wish. In addition, it is good to *Rubify* in other works, etc.

TINCTURA MARTIS

Take iron filings and put them into distilled

vinegar. To every pound of vinegar add ⚗✳. Set it for six weeks in the *Balneum* or in horse dung. Then take it out, pour the vinegar off, and add other vinegar

together with the addition of ✳, as before. Put it in the *Balneum* for six weeks, and pour all your tinged vinegar together. Distill it all together in the *Balneum* per alembicum. A red tincture will stay at the bottom, like blood. Take that out, sublimate it ten or twelve

times with ✳; then dissolve it in the cellar on a stone; coagulate it again *Per Alembicum In Balneo*; or you may previously dissolve it in water of the Hand, and extract the *Aqua fort in Balneo* and from the sand, but then you must not sublimate. You will be left with a redness more beautiful than gold. This is called *Crocus Martis*. With it you can do wonderful things, so that it is much better than a landscape; since by dissolving it in *AF.* and again congealing it, you can make it so subtle that it reaches very great power; for with that one can

congeal ☿ to a beautiful redness; in addition, one can Rubify with it everything that is white.

TINCTURE OF ⟁̄

The art of extracting the *Qu. Ess.* of ⟁̄ , *Auri-pigment*, *Calcined Ochra*, *Atrament*, red *Arsenicum*, etc., is all one procedure, one art, one work, as follows:

If you happen to have some stuff and wish to extract the tincture from it, powder it impalpably; pour on it old pure urine; let it boil till the urine is colored. Then pour it off, and again pour fresh urine on it till all tincture is out. After this, evaporate all the tinged urine. Pour vinegar on that which stays at the bottom. Extract again all the tincture. What then stays at the bottom is of no use, because it is the saltiness of the urine.

Draw the colored vinegar off *Per Balneum*, and the most beautiful redness will stay at the bottom. It shines more than ☉, and it seems to be a heavenly color. Dissolve this tincture in the water of the Hand; distill same, like the previous tincture, by dissolving, distilling, congealing, and preparing it to its highest degree and power. Or you may sublimate through ✳ as often as you wish, dissolve in the cellar on a marble slab, and bring it to its greatest power. This was the first manner of the philosophers, and it was a long way. With these tinctures you can *Rubify* all things, yes, all metals, also all stones, crystals, glass, and other things, which I do not reveal for certain reasons.

248

TINCTURE FROM ANTIMONIUM

Take *Antimonium* from the mines, pulverize it impalpably; pour on it distilled vinegar in a glass or a stone jar. Put it for six weeks in horse dung or in the *Balneum* — but the horse dung must be renewed every week, the longer the better. Subsequently, put it into a retort with an *Alembic*, draw the vinegar over with boiling water; then drive it in the sand, and it will sublimate into the blessed *Qu. Ess.* and come down from the spout of the helm like red drops of blood. Collect those *A Part* and mix them with the Key, which must be prepared so dry that it is like dust.

Now distill Per Balneum over a gentle fire, so that you can easily suffer your hand to be in it. Let it stand thus, distilling for twenty—one days, or till there is no more moistness. Congeal the Qu. Ess. ☿ ii and, in addition, all moist spirits, so that you can sublimate them. After this, take it out of the *Balneum* and set it in sand, in order to sublimate the red Qu. Ess. ☿ ii. When there comes a heavenly, incombustible redness, though not fixed, that is how it should be.

Now take it out carefully, put it in the *Balneum* to putrefy for eighteen days, after which mix it with the dry Key, dry like dust, as you have been taught before. Thus you can dissolve and congeal to reach such great

virtues and powers as are worth a kingdom. When you have
done it, add a great deal of gold, as I taught you to
make before; dissolve them together in water of the Hand,
distill them in the *Balneum* and afterwards in ashes, as I
have instructed you before. Then the Qu. Ess. ☿ ii is
conjoined to the ☉ and can never again be separated,
either now or in eternity. *Deo Gratias.*

A PHILOSOPHERS' STONE

R. Fine ☽al ℥ j, dissolve it in conunon *AF*. Then take

Mercury Sublimatus ℥ iiij; put him in a glass pot, set
it on warm sand and let him get warm. Now imbibe him with
the *AF*. in which ☽ is dissolved, till he has imbibed all
the *AF*. Then let it cool down. Pound this Mercury quite
fine on a hard stone; let him dissolve on it of his own.
After this, coagulate him again on a small fire in a
glass; and pound him again as before; and dissolve him as

before, seven times. Of this Elixir pour ℥ j on ℥ xxxj
prepared, well flowing copper, and you will get fine
silver in all alloys.

HOW TO MAKE VENUS FIXED

Take soap and dry it till it no longer smokes. Then
give a hot fire till it turns into chalk (calx), When you

250

calcine it, stopper it above with a stone. Add to it as much verdigris and close it in a crucible; lute it well, set it in the fire for two hours and melt it. Then you have fixed and supple ♀.

MERCURY CHANGES ALL METALLIC CORPORA INTO SILVER WITH ARSENICUM, AND INTO GOLD WITH SULPHUR

Take a thick cloth, bind Mercury into it, hang him over a pot containing sulphur; then close it and heat below it. It will congeal red. But if this Mercury is pounded with *Sal Alcali* and *Cerussa*, he turns white. Instead, if he is pounded with water of Crocus Martis, ✳ and *Atrament*, he nevertheless stays red, and one can also congeal him in a closed vessel with ⯛.

With Saturn Mercury is congealed in white works. He has to be closed in a tube (or: pipe) or another vessel, luted with *Lutum Sapientiae*. This has to be thrown on melted *Saturnus* and also kept in the fire; thus *Saturnus* will be congealed by the smoke.

HOW TO CHEW THE CORPORA TO AN AMALGAMA WITH MERCURY, THAT IS, TO GIVE TO MERCURY ALL CORPORA TO EAT, EXCEPT ♂

Take any Corpus that you wish to melt. Add to it half as much Mercury and pour it. It will become brittle.

Pulverize and wash it with vinegar and salt as long as blackness goes off it. Afterwards, dry it at the sun. Then it is to be cleansed with a water whose taste is between sour and bitter, with a gentle fire or the heat of the sun, for a whole day. After this, dissolve it; when it is dissolved, *Incerite* it; finally, cleanse it with *Arsenicum* that you have whitened; then dissolve it again; after this, congeal it. This *Materia* will tinge the bodies of all men in a wonderful way.

THE PHILOSOPHIC FURNACE, CALLED ATHANOR, OR THE WARM STUPHA OF THE PHILOSOPHERS

First build a little wall at the bottom with good glue (or: lute), or with soil prepared for this purpose - one foot long and high. Upon it put a narrow iron grit, so that the ashes of the wood or the coal can fall through. But under the grit there must be a little door, which can be shut as one wishes, so as to remove the ashes. Then, above the grit of the furnace build somewhat higher, a little wall, one foot long and wide, with a small door, and air holes at the four corners. Above, make a whole floor of baked soil, without holes, so that the furnace is well covered, as this illustration shows. On this floor raise the wall; but first put on the floor, four iron supports like a tripod. On top there has to be a vessel. These supports together with the vessel must not touch the wall of the furnace. In addition, there has to be another tripod inside the vessel or test, as also a small vessel upon which one could put a glass or other bowl. This small vessel must be made of wood.

If then one wishes to dissolve the *Spiritus* or other things, water has to be put into the lower and upper vessels. The glass has to be left open, to allow the humid, subtle air to get inside. Now put a gentle little fire under the furnace, because it must not be warmer than it is usually in a warm room. Thus the matter will be dissolved by the steam of the water

But if you wish to congeal, calcine, or fix, you must put no water into the vessels but keep your work quite dry with the aforementioned heat. That is why the furnace must have a lid or cover above, which must be so wide that you can take it off if necessary. In addition, there should be a little door in the cover, so that you can put your hand inside and test the warmth, just as with the same heat all humid spirits are fixed. Will you please have a good look at the illustration.

Aside from this, there is still another furnace in which many kinds of things can simultaneously be cooked or done, as in ashes or in sand, in *Balneo*, or in *Ventre Equino* (the horses abdomen). This means, that if someone would like to putrefy something which would otherwise have to be put in horse's dung, one would put hay or dung into the cupel and water it somewhat. Then it is just as good as if it were standing in horse's dung.

IT NOW FOLLOWS HOW ONE IS TO MAKE OR USE THE FURNACE

First, the lower part has to be made, so as to get the ashes. Then follows the grit which must be just in the center, a good bit narrower than the furnace. The furnace has to be made higher yet, but not narrower than it is below. On one side there has to be a small door, so as to allow the hanging lamp and, below, the coal to pass through. Afterwards the furnace must be made higher again and wider, with a thin wall. A few air holes must be made on one side, which can be opened and closed at will. After this, it must be closed above and vaulted, well—fitting, with four round air holes with a lid on each of them. They must be separated by four equal little walls.

After this, the furnace must be made higher on the sides; it must have four doors on the four sides that close tightly, so that, if one wishes to remove the lid from a hole, one can put one's hand through it. When all this has been done, take four copper kettles, or other vessels burnt of clay, or something else. They must be walled in on the four parts of the furnace, one on each side. Next to each cupel, there must be an airhole with a well-fitting stopper in order to regulate the heat. In

the first cupel or kettle there must be sand to make
fixation in it ("figieren"). In the second there must be
dung or moist hay, to putrefy in it. In the third, one
can keep ashes, to distill on them. In the fourth, should
be the water bath *Mariae*, also to distill and putrefy.

When all this is ready, one can put coal on the grit,
or a hanging lamp. The four lids have to all be taken
off; or, if one wishes one to be hotter than the other,
one should remove that one lid and shut the little door
tightly. One must also have pointed and high lids, burnt
of clay, on the glasses in which one wishes to putrefy
and fix. They must cover the glasses all around and close
them tightly, and one must be able to take them off, as
may be seen by the illustration.

OPERATIO TARTARI, OR THE WORK OF TARTAR

This is the way to prepare *Tartarum*. It is very good
and has the four elements in it, but not like other
things, Species and herbs.

In all the things that God Almighty has created out
of the four elements, the elements are befouled and
impure, and they cannot be purified otherwise than by
distilling, sublimating, calcining, cementing, etc. In
that way the elements can be cleansed with hard work. One
cannot find anything in the world that is like tartar,
because God has not left any impurity in it, since it
separates from wine of its own, like water from fire; and
tartar does not take on any impurity from wine, just as

water does not from fire; but if there is any impurity in the water, it will leave it in the fire, so that the fire may burn and destroy it. For fire consumes every impurity, and all impurities of the elements must be purified by fire. Nevertheless, fire is in itself frail in that it is not fixed; for when God corrupted the four elements, it also corrupted the fire. That is why the fire is not fixed.

Tartarum, however, is a fire without air and without a flame, fixed and pure. That is why no one can separate the element from *Tartarum*, since it is a glorified *Corpus*. Nothing can destroy *Tartarum*, nor can fire burn it. Tartar is the fixed fire of which we often write in our books. Fire has the power to purify all corrupted elements and to burn and consume all impurities, also to make all other elements fixed. What will this one (tartar) not do when it gets into the impure elements, since it is more than superior to fire? Verily, I am telling you that *Tartarum*, when it is prepared, has the power to make all unfixed things fixed, and in it are hidden all things which it would take too long to describe. Yet one cannot accomplish any perfect work in the Art without *Tartarum*; thus the Hand cannot be prepared without *Tartarum*; for if there were no *Tartarum*, the Art would be false. Consequently, *Tartarum* is the Master of *Alchymia*. That is why the masters consider it their fixed fire which burns everything that is not fixed.

Take good white tartar of good Rhine wine. Wash the powder clean till the water runs off it. This has to be done with pure spring-water. Then dry it at the sun, pound it small on a stone with good distilled wine vinegar, and dry it at the sun. Then moisten it on a stone with fresh vinegar to a thick pap. Now dry it again. Repeat this at least ten times or more. Put this *Tartarum* thus imbibed into a strong suitable vessel and close it firmly, a good hand's breadth.

Now take another, still larger vessel, fill it with living chalk (quicklime). Into it put the vessel with the tartar; after this, fill the large vessel completely with living chalk; also around the tartar, so that it lies in the center like an egg yolk in an egg. Lute tightly as much as you can, a large hand's breadth's, and let everything gently dry of itself.

Make a big hole in your hearth in which you have a fire every day; fill it with living chalk, put the vessel in it, and fill it all around with living chalk; but leave the bilge of the vessel free. On this your fire must burn for eight weeks; keep it hot day and night, afterwards in a glow. The more it is calcined, the better. After this, remove the small vessel from the chalk, break it open, and pour the *Tartarum* into a large quantity of distilled vinegar, let it drop or become pure during three days. Pour the pure off from above and into another vessel. Pour other vinegar on the feces, boil it and let it become clear as before; pour it to the other vinegar. Repeat this till nothing comes off any more.

Throw the feces away and *Evaporate* your vinegar; congeal
all the salt till dry; again dissolve it in a large
quantity of vinegar, as before. Let it boil once more;
let it settle again, and pour the pure off above from the
fecibus; add other vinegar, and repeat this till no more
feces remain.

Now put it into an open crucible, let it glow for
twenty-four hours in a furnace; then dissolve it again,
and let it stand thus dissolved for three hours in order
to see if there are any more feces. Should it have feces,
they must again be clarified off, till no more feces
occur; then congeal it till it is dry. After this, imbibe
Tartarum on a glass with *Aqua Vitae*, to a thin pap. Dry
it at the sun or on hot ashes in a glass; imbibe and dry
it again. Do this one hundred times or more; the more the
better, and the more it gets purged. Thereafter dry it in
a glass which can stand the fire; lute it tightly below,
put it uncovered on a furnace and let it melt. When it is
molten, break the glass open. It will congeal at the air
as clear as glass, and melt near fire like butter. This
is the fire of the wise philosophers. It is their fixed
fire, of which they write in a veiled way which the
simpletons do not understand. (Sophic Fire)

When it is now thus prepared, it is a precious
treasure, for with it all volatile matters can be fixed.

Yes, put ☿ into a crucible with a little of this
Tartarum upon it, put it in the fire and let it go, and

it will turn into fixed ☿. In various small works one
has to prepare it in this way; because, whoever works
with it when it is not prepared, will not reach any

Perfection, although he may accomplish a semblance of ☽
and ☉ . Also, those who thus work in unprepared *Sales*
cheat themselves and all those who believe them.

SOL AND MERCURY

Here begins a most glorious work of ☉ and ☿; how to
conjoin and unite them; also how to fix them together
into the very best gold.

First you must make an *Aquam Mercurialem*. I will,
however, advise all people against this Art, if they do

not know how to deal with ☿, nor can prepare the water
described hereafter.

Let them keep their hands off ☿, for they would fare
badly. This water is of two kinds, one to the Red and the
other to the White. Take 4 Cologne quarts of good wine

vinegar; potash, 1 lb, quicklime, weed-ashes, ♀ well
calcined in a potter's furnace and afterwards pulverized,
each 2 lbs. First pour the wine vinegar on the lime, let
it stand on it, and stir it with a stick seven or eight
times a day. After this, filter the vinegar and pour it
on the weed—ashes in a glazed pot. Let it boil for a

quarter of an hour, then cool down. Filter it again and pour it on the Tartarum; boil and filter as before.

Now take 6 lbs of green, common celandine; chop it finely, put it into 2 quarts of sharp wine vinegar, boil it down to half, then filter it and mix it with the other aforementioned water. Then take 1 lb *Aqua Vitae*; coarse salt that has not boiled away, 1 lb. Let them boil up together, but be careful not to put your matter in vessels coated with lead. Into this water thus boiled, put eight or ten ox galls, or galls of other animals. Mix them, and put this matter into a glass pot; put it for six weeks in horse dung, and it is ready. The older it becomes, the better it is. With this water one accomplishes wonders. It is called *Sucus Mercurii*, and it is very good piece in this Art.

Take, therefore, good ☿ from the mines, not made from ♄, ♃ or other metals. Take as much as you wish, grind it well with salt and vinegar in a stone mortar. Do it twelve times, till no more impurity comes off it; make a *Corpus* of it, put it into a phial and add as much beautiful, white salt. Mix well, put it on the furnace in sand for three weeks with a gentle fire; then put it into a stone mortar, pour the aforementioned water on it, grind it with it till no more impurity comes off it; put it back again into ashes or sand, pour your prepared water three finger's breadth above it, stopper the phial, and give it a good fire for six days and nights. Then put it back into the mortar and grind it again with the same water, as before; put it again into a phial, and again pour water on it, and give fire for six days and nights, somewhat stronger than before. Repeat this six times, and you must increase your fire each time every six days.

After this, calcine 4 or 5 lbs of *Vitriol*; also calcine an equal amount of oger ; to each pound add ℥ i bloodstone haematite to calcine with the other. Through this sublimate 5 times; then he is prepared. Now take ☉ which has been cast through ♂ ℥ j; prepared Mercury ℥ viii, amalgamate it, put it into a well luted phial, put it on its side in sand, and at first give a gentle fire. Increase your fire daily, till ☿ no longer rises; and turn the phial over everyday four or five times, the more the better. At the end, when ☿ is quite dead, you will find a brown powder. Now your work is accomplished. Now remove this powder from the phial for the praise and thanks to God Almighty. Put 1 part on 4 of fine ☽, let them well incorporate together. This work is certain and without troubles. That is why you must use it in the name of God.

HOW TO MAKE SAL URINAE

Take several large, stone crocks, 10 or 12 quarts capacity, fill them with old, pure urine, five or six weeks old. Put an alembic with two spouts on, and to each spout, a large *Recipient*; distill on the fire whatever you can distill over. Then a great deal of blackness will stay in the jar. Take that out and calcine it for two or three hours. After this, remove it from the fire and dissolve it in common distilled water; let the feces drop, and pour the pure off the fecibus above. Put this pure water back on the fire and boil it till there forms a skin above. Then put it into a cool cellar, and a clear salt will crystallize. Take that out and boil the water down again till a skin forms.

Now take all the crystallized salt and dry it in an earthenware pan till it powders (or: dusts). Now heat it moderately without melting it in a low vessel, and dissolve it again in common distilled water. Let it boil for a quarter of an hour, remove it from the fire, let the feces settle down, pour the pure off above when it is still warm. Then boil the pure water down again till a skin forms. Take it again into the cellar and let it crystallize. When it is crystallized, take it out, and boil the water down still more, and let it crystallize. Then dry all the crystallized matter in an earthenware pan as before, till it dusts. Keep this till I teach you how to use it.

Now take all the distilled urine and, if there is
still some impurity in it of oil or yellow greasiness,
skim it off above with a spoon and afterwards with a
feather. Make it pure and clean in this way; then put it
into the stone jar with a helm on it; set it in sand or
ashes, and distill all that can be distilled. Let it
nicely glow for a while and afterwards cool down. Remove
the feces and throw them away, for they are good for
nothing. Repeat this distillation till everything goes
over pure, without leaving any feces at the bottom of the
jar or retort.

Take that which you have thus distilled out of the
fire, put it in the *Balneum* to distill. Some impurity
will stay at the bottom; throw it away. Grind this
Distillation in the *Balneum* till no more *Feces* stay at
the bottom of the jar. When everything has cleanly gone
over, it is done.

Now take the salt which I told you to keep above. Put
it into a large *Recipient*, and pour on it the pure liquor
which you have distilled in the *Balneum*. When you have
poured the water on it, stopper the *Recipient* with a
cork, and put it on the furnace in a vessel with ashes.
Now light a fire in the furnace, so that the ashes will
become hot. Keep the *Recipient* standing there for four or
five days, or till all the salt has been dissolved into
pure water without *Feces* at the bottom. Then it is ready
and done. Then it is urine as it was before, but now it
is deprived of its coarseness and has become subtle, yes,
much subtler than one could believe. I am telling you in

unadulterated *Charity* of God that one can accomplish wonders in our Art with this urine thus prepared, yes, more than one might believe.

Add to it 6 quarts of sharp, distilled vinegar, and 6 Mass of *Aqua Vitae*, and 1 lb of prepared common salt; also, 1 lb of calcined white tartar, and 1½ lb of ✳. Mix all these things together, and dissolve the matter into a pure, clear water, without any feces at the bottom. I swear to you by God who has created and made me, that no greater secret has ever come into the world. For this water thus prepared, turns all *Calces* and *Corpora* into their first nature, that is, into ☿. With this water one can draw the *Qu. Ess.* from calcined ♁, which is much better than all the treasures of the earth. In addition, you can extract with it the *Qu. Ess. Solis* and *Lunae*, and, furthermore, of all mineral things, and of everything there is in the world. One does so many wonderful things with this water that it is unbelievable; nor is it permissible to disclose it on account of the evil which might ensue.

Understand, however, this booklet well at bottom, and you will know what wonderful things one can perform with this water. And even if I did my very best, I could not express one thousandth of its secret. Know also that one can use the water like earth, for it does not diminish although it were used ten or twelve times. For you may purify and rectify it again, and it will be just as good.

Now I will teach you how to extract the tinctures required for this work, Red as well as White; for the
extraction of White and Red is all one and is due to one

mastery. Take, therefore, ⚥, *Auripigment*, *Atrament*,
Ochre, *Cerussa*, *Minium* (red lead) or the like, out of
which you wish to extract the tincture. Powder it
impalpably, and grind it like soap on a stone with good,

distilled vinegar, each time adding ℥ j of rectified
salt. Put it into a big *Recipient*, set it into ashes or
sand, pour on it some of the clarified urine which you
have prepared, 1 part; and 1 part of wine vinegar, so
that the Recipient becomes half full. Stopper it above
with a cork, and shake it well, so that the moisture gets
well mixed with the powder. Put the glass back into the
furnace, and when it is warm, remove the cork or stopper
and give it some air; otherwise the glass would burst.
Toss the *Receptacul* between your hands ten or twelve
times a day, and let it thus stand in the warmth till the
Liquor is nicely colored. Then let the glass cool down
and the *Feces* drop; take another large *Recipient* that is
clean. In it pour the colored *Liquor* off from the
Fecibus, and take good care not to take any *Feces* over.
Stopper the Recipient and put it aside. Then take again
fresh urine with *Ana* distilled vinegar, pour it upon the
Feces in the receiver, as before, half full. Shake it
with your hands as you did then; and when the *Liquor* is
well mixed with the sediment, put the glass again on the

266

furnace in ashes or sand. Heat as before, and when the *Liquor* has again been colored, let the glass again cool down, pouring off and proceeding as before. Repeat this till the *Liquor* is no longer colored by the *Feces*, and then you have got all the *Tincture* or *Qu. Ess*. The *Feces* can be thrown away, but they still contain the element of earth. You may extract and use that as you wish.

Now take the glass containing the colored moisture, set it in ashes or sand, and distill the moisture off till a skin appears on top. Now remove the helm, pour it into a large glass pot which must be wide above. Lute a helm on top of it, put this pot on the same furnace, and draw all humidity over. Thus the *Qu. Ess*. or *Tincture* of the thing you took to make it will stay *In Fundo*, be it Red or White. If it is a white *Subjectum*, you will find a white tincture, whiter than snow; but if it is red, it will shine like ☉, just as the sun shines above ♀. And in this way you can extract, sublimate, the tincture of ☿ *Sublimatus* to the Red or the White. In the same manner you can extract the tincture of *Qu. Ess*. from iron or copper filings, verdigris or burnt (calcined) ♀, cinnabar, *Cerussa* or *Minium*, or from calcined ♃, also from *Calx Solis* or *Lunae*, item from ♄. In addition, you can add the tincture, which you have thus extracted from sublimation to the Red or the White, to ☉ or ☽, when they are dissolved, and put them in *Putrefactionem*. Thus

267

they can be joined to the said *Medicine*, and the color will become all the more beautiful during *Projection*.

Know, however, that whenever you wish to extract some-

thing you must each time add $\mathcal{Z}$ j to the urine and distilled vinegar, which must be rectified of its *Terrestriality* by dissolving and again congealing before you pour it on the ground *Materia* in the *Recipients*. Know, in addition, that there is a great deal of secrecy in the *Extraction* of this tincture, more than one could believe. For with these tinctures you can make *Cements*, in whatever *Cement* you wish to *Cement*, or accomplish other wonderful things by *Cementing*. Item, you can make *Aquas Fortes* with these tinctures, which are as red as blood and shine like a ruby. With these *Aquas Fortes* you can do wonderful things during *Solution*, to disclose which I have not got permission.

NOW FOLLOW SOME ANNOTATIONS CONCERNING VARIOUS TERMINI USED IN THE PREPARATIONS

LIQUEFACTIO means making soft, and it is the root of all things.

EXALTATIO of the Spirit is LIQUEFACTIO of the bodies.

The SALTS of the bodies are made in many different ways.

SOLUTIO means melting, or also stamping (or: pounding).

CALCINATIO means making chalk (calx). It is done in sand, with strong fire, so as to draw the foul *Sulphur* from a thing; then, when such is consumed, the corpus remains pure in the chalk. Thereupon comes INCERATIO.

INCERATIO means making the chalks subtle or fine, so that the humidity may pass all the better through the Corpora.

INCERATIO is nothing but grinding and imbibing till the Materia turns into wax, and such may easily be dissolved or melted with a small or gentle fire.

DISSOLUTIO means delivery, release. It is done in the bodies that are calcined, and in the following manner: Put the matter that you wish to dissolve into a glass, and stopper it well above. Cover it with moist earth, and put on it horse dung well sprinkled with water. Let it stand thus for seven days. On the eighth day it will be dissolved, and this is DISSOLUTIO by heat. But DISSOLUTIO of cold humidity consists in making an edge ("kante") and standing a glazed vessel on it. Complete this with a little water on the bottom. Let the glass hang above the water, but in such a way that it does not touch the water. Cover the glass vessel with a lid, but put wax on the glass, and sand on the wax. Let it stand thus for one day and one night, and it will dissolve.

COAGULATIO means making hard, and it is done in this way: Take the aforementioned glass with the Matter that is dissolved, put it into a test with strained ashes, and light a fire under the test for ten hours, or till the humidity has gone out. Then it will look like miniun; but it is white, like a very white camphor.

RUBIFACTIO SALTS ARMENIACI (AMMONICI): Take ✳, *Crocus* ♂ *Ana*, well pulverized together. Put this into a glass and pour on it good distilled wine vinegar, three fingers breadth above it. Let it stand thus for one day and one night, always stirring well. Then allow it to dry gently over a mild fire. After this, put the Crocus ♂ and the ✳ into a Sublimatorium, and sublimate them. It will descend, red like blood, even if the vinegar is not distilled. This actually is of no consequence, it is just as good.

RUBIFACTIO VITRIOLI ROMANI: Graduate it, and calcirie it in a strong fire; it will become red like blood.

HERE FOLLOW FURTHER TRACTATES OF ISAAC HOLLANDUS

The Art of Alchemy consists in three things, that is, in our Stone. That is the free art of the ancients and their successors, who are to discover this free art through Science or *Practice*; or to whomever the Holy Ghost give it, or upon whom He confers it through His illumination; and blessed is he who possesses this free art and applies it wisely for the honor of God and the pressing need of his fellow man.

The other kind of Alchemy is the elixir which is prepared, as the ancients taught, according to the Hand of the Philosophers. Those who have this and understand it well may also be called blessed.

The third kind is the *Ixir*, and it is also an art of the old and wise Masters of the Hand, and he who knows how to prepare it, as the fathers prepared it and have left it to us, will likewise rule in this world in joys.

The Art of Alchemy has still many more daughters, branches and roots which spring from these three trees of which I have just spoken, such as: Some labor with hard work in the Calcination of the bodies, to wash them and make them pure and clean. Others labor with *amalgamationibus*; others with *albination*, *cementation*, *augmentation*, and *rubification*. Others make *Salia* of the bodies or the metals. Others make *Olea* and other works of the bodies in the fire. Others with *aquis fortibus*, others with *salia* upon the *corpora*; and so on in many different ways. And everything is good if it is done in the right way as the forefathers have taught; but all this is attained with great effort. Once perfected, painters and goldsmiths could not counterfeit. As the *materia* becomes riper and stronger, so there occurs many a change. Before you reach the grey-black powder, your materia will turn a beautiful yellow like ashes of wood or peat. All this happens with a small regimen of fire. When you see the grey-black powder, rejoice, for under the blackness the whiteness is hidden.

After this, let the *materia* stand in this regimen of
fire for a long time and watch if it retains this color,
or if it becomes whiter and clearer. If it turns whiter
or paler, keep the same regimen of fire; but if you do
not see or sense that the color is vanishing somewhat, or
that it changes, increase the fire a little, until you
see that the color turns somewhat paler or whiter. Then
let it stand for a long time in the same regimen of fire.
If the color does not get stronger, increase the fire
somewhat, and as soon as you see that your materia
becomes whiter, maintain this regimen of fire and be
careful not to make the fire too strong. Follow this
regimen of fire, increasing it each time just a little
and never too much. Do this until your matter is white,
yes, whiter than snow. Then rejoice, dear children, and
be assured that under the White the beautiful Red is
covered and hidden.

Morenius says: When Christ lay in the tomb and was to
make his resurrection from there, so that after such a
resurrection there should be a *gloriosum corpus*, which
was to live in all eternity and was to be crowned with a
red diadem and be king over all his generations, and all
his enemies were to make peace with him, and he was to
remain king throughout all eternity.

Now then, you must understand that this white materia
or earth which you have is only an earth that has lost
its humidity and is not yet good for anything. That is

why you should know that many mistakes occur in this Art, because there are many who dare to undertake to confect the Philosopher's Stone and achieve this stage with a good regimen of the fire. They labor to coagulate or fix their Stone to the White or the Red; and when the *materia* or Stone is fixed, they believe they can make *projection* and throw thereon raw *Mercurius* or other *imperfecta metalla*. They work in vain, however, and fall into despair saying that the Art is impossible. It is true, that to <u>them</u> it is impossible, because they have an earth which has lost its humidity, just as *Geber* says: Spirits which have lost their humidity because of many *sublimations* and *fixationes* are good for nothing as long as they are dry as this one. Ignorant men do not comprehend it, for when they have made their Stone and joined it well, so that it has its right color as is necessary and later has been made subtle and volatile again, it gives quite another *ingress* or *projection*. But they do not understand the words of the sages. They may well know that the Stone is to be made in such a way as has been related above, and they do indeed make it in the manner prescribed; but, they stop with their work just when they should start working all the more and thus remain in their foolish mistaken ways.

You must know dear *Filii*, that I wish to reveal here the truest secret there is in the Art that I know and has been disclosed to me. Therefore, I entreat you, by the living Son of God, that you do not divulge this secret to anyone except your own sons, provided you believe that they have the love of God and that, in addition, your

273

soul and mine too, will not be damned on account of it
due to the great tribulations that might arise from it.
Open your eyes and ears, and hear the great holiness that
is in nature. In the Great Work of which we are here
speaking, both to the White and to the Red, you can make
and perfect, in one vessel, and one furnace, all *Lapides
Philosophicos* in whatever manner they are composed.

Take good heed of what I am telling you. If you take
☽ and *Mercurius* without adding ☉and put them together
in this manner, you will soon be able to make the *Lapis*
from them, to the White or to the Red, in one cask and in
one furnace. Now then, someone might ask why one puts ☉
and ☽ together in this work. This is done because is
fixed, and the work will thereby become all the shorter.
If your ☽ were fixed in this work, the Philosopher's
Stone would be ready to make to the White. However, ☽ is
not fixed, therefore much time must be spent in cooking
in order to fix it, because ☽ must be fixed before it
can fix *Mercurius*. That is the reason why ☉ and ☽ are
put together in the work. You can also make both stones
of *Mercurius* and ☉ of alone, and that would be done
faster than with ☽ and ☉ together, because ☉ is
fixed and it will therefore fix (coagulate) *Mercurius*
quickly.

Someone might now ask what would be the result if one were to take ☽ alone with *Mercurius*. Could the red Stone also be prepared from them? Understand it thus: ☽ is red in its innermost, just as it is white outsides; for in all white things in which the four elements are, there is redness within covered outside by the white. ☽ is cold and humid, just as is *Mercurius* and ☽ when still raw and unfixed and coagulated together. That is why it is white outside and red inside.

When now ☽ is alone in the work with *Mercurius* it is cooked completely and coagulated with a good regimen of the △. And when it has become fixed during the work, it coagulates *Mercurius* together with it, and it turns into a white Stone of the philosophers by means of an increase of the fire and long cooking. Then the white stone is colored red, and its *tincture* emerges to the outside and the white recedes within.

Ponder this well, dear children, that which I have said and that which I will still relate, for it is all together necessary. It is the *secretum* of all *opera*; there have been some ignorant men who, after they made their Stone to the White and to the Red, saw that the Stone had no *ingress*. Then they dissolved the Stone and coagulated it again. They did this 20 or 30 times, hoping thereby to make the Stone fusible, so that it should have ingress. However, they did not succeed, even should they have dissolved and coagulated until Doomsday!. The Stone was bound to stay as it had been previously.

There have also been others who have drawn an oieum out of ☿̄. With it they rubbed the Stone on marble, dried it in a glass, and *imbibed* it again so long and so much until the Stone became liquid like wax, and gave an *ingress*. Following this, they threw it upon *Mercurius* which had been made red hot; and as soon as *Mercurius* glowed, it flew away, and the *oleum* ☿̄ followed it, and the powder of the Stone stayed in the crucible, as it had done before they *imbibed* it. But this was due to the fact that the *oleum* had not been coagulated with the Stone. The reason is as follows: If you had thus *imbibed* the Stone and put it in a glass over a fire, the *oleum* would dry along with the Stone. But if you were to give it strong beat, the *oleum* would fly away altogether on account of the great heat. That is why one cannot coagulate together with the Stone. Thus the ignorant remained mistaken.

Now I will teach you, my child, how to make the Stone fusible and capable of giving an *ingress*. Until this hour, this has never been revealed. Therefore, dear *Filii*, do not divulge this *SECRET*, by the love of GOD, your soul and my soul.

After your Stone has turned white by means of a good regimen of the fire, as I have taught you before, and you wish to keep your Stone white, it is up to you. However, if you wish, to wait for the time that it turns Red, you must leave it longer in the furnace, increasing the fire considerably. When you see that it begins to get yellow like gum mastic, do not increase the fire. Let it stand for 8 or 9 days in such a heat, and observe if the Stone has turned somewhat more yellow. If it has remained the same hue, increase the fire considerably. When it begins to take on the color of Saffron, again let it remain in this heat for 8 or 9 days.

Proceed in this way, through the regimen of fire, until you see the Stone assuming its perfect redness. It will be like glowing Gold standing in the fire. It appears to be more of a heavenly than of an earthly color. In this way, the stone must be cooked with a strong fire. With only a small fire, its *tincture* and Sulphur do not come through. It is a red tincture, and before it reaches its perfect redness, it must stand for 41 days.

Know that if the Stone were fluid, its redness could not be brought out. This is because if it were glowing hot, it would melt and even penetrate through the glass! Thus, everything would be lost. Finally, it must glow for three days. Concerning this, note should be taken that the Stone must first of all have been made before it is made fusible. That the ignorant cannot understand or take note of, because they do not understand nature. In this

277

way, both the white and the red Stone must be made before
they are made fluid and subtle, as you may well
understand with your intelligence.

NOW HEAR MY SONS, THE GREAT SECRETUM

Which is in the Art and which has never been put
down in writing except by myself alone: That is — how one
is to prepare the two Stones and make them fusible so as
to make projection with them.

Take your Stone and pour over it clear, clean
paradise-water. Join it to the water and set it in the
prison and close it well. Now it will rise toward heaven
during one rotation of the moon, and turn into a dew, and
again fall down, drop-by-drop, according to the teachings
of the Masters. It will moisten the earth, so that it
should bring forth flowers of various colors. When these
flowers appear, your Stone will rise from the dead to a
body; and all enemies shall make peace with it; and the
tempestas which were before, will be over. And it has
overcome the darkness, and the *eclipis* of the sun and
moon, and shall forthwith remain a king of all its
generations, and he shall not lose his dominion in all
eternity but remain the Rex Glooiae.

Remove the Stone, either red or white, from the cask
and put it into a stone mortar, pour upon it a goodly
amount of ♀ *purgati* - as I have previously taught you to
prepare - and mix them with a wooden pestle for an entire

278

day without ceasing. After this, return it to the glass, set it back *in furnum philosophorum* or *in tripodem*, give

it a good fire, such as will keep ♄ in flux. Lute the mouth of the little glass that has the Stone inside and

keep it in this heat until all the ☿ is dead. That happens soon enough, in about 40 or 50 days. Then, the Stone of its own accord, draws its spirit into its nature, because like attracts like, and all rejoice with their likes.

When now the ☿ is dead, increase your fire just a little, so that the *materia* turn white. When it has

become a white Stone, take it out, heat a ♀ sheet or plate until it glows. Put one grain of the Stone thereon. Now observe if the Stone becomes fluid and if it has an *ingress* so that it *tinges* the *lamina* and goes through them like an *oleum* goes through dry leather, and makes the sheet white like fine silver. If it indeed does that, it is ready. If not, pour once again clean paradise-water

over it, just as has been previously taught. To 1 ℥ lapis take 4 lot paradise—water, as often as you do pour paradise water on it. Do this until the Stone is fluid and has an *ingress* such as you wish it to have. If it is the red Stone, and you have poured paradise-water on it, let it stand in such heat as would keep it in flux, without glowing, till it turns red again. It takes much longer than with the white Stone. Test it also like the white was tested.

But you must take not of the following: If you wish to prepare the stone in order to make *projection* with it upon ☿, it must be made as fusible as wax. This must be done with care, so that the stone does not penetrate through the glass. My advise, however, is that you make it as fluid so that it is absolutely glowing before it melts. It is the white stone to which this applies.

Therefore, make *projection* with it upon ♃; but you must not make the red Stone more fluid than that it may well glow but not blow. For when the paradise-water in it has died and is fixed, it must stand in the furnace and glow for 40 days before the paradise—water comes out red. When the redness is outsides, you must increase the fire so that the Stone glows steadily, so much so that one can see that it is glowing and no more. Let it stand thus for three days then let it cool down, and give thanks to God that your ✡ Stone is perfect. (*Note: "glow" means a red or white heat, to anneal*).

Dear *Filii*, you must have moderation in all your works, and especially in the process of making your stone fluid. Do take great care not to make it too fluid or it will go through the glass as has been previously taught. The red Stone must be made even less fluid, or you cannot add to it the tincture of the paradise—water; for you must know that you can make everything in the world supple and fusible, if it is sublimated together with it so that it does not leave it. This is called *ceratio*, and *ceratio* is nothing but the process of making a thing that is hard and not fusible, fluid, so that it gets an *ingress*. This was first discovered by the ancients. After they searched for a long time to discover how to achieve the Stone to the White and the Red, they found it was not useable when they desired to make *projection*, because the Stone did not melt and stayed at the bottom as a powder or earth. Thus they noticed that they lacked nothing but *ceratio*, which would give it *ingress*. They looked in many different things and yet did not find it, except in

Sulphur and *aurip.*, and especially in ☿.

Likewise you must know that the oil in all things in the world will separate from its earth in fire, except that of *mineris* and *metallie*, because their oils stay with the earth in the fire and do not separate from them. And when they separate, the earth rises together with them, because these oils cannot be separated from the earth, as can be done with other matter. So they now realized that if they wished to follow nature, they required such oils, to *incerite* and liquify their spirit and dry earth. They found such in *Sulphur* and in

auripig., but ten times more in ☿.

In this way the art *cerationis* was discovered. They liquefied their stone as they wished, and it did for them whatever they desired of the Art. They thereby liquefied whatever they wished, they *sublimated* the spirits through hot matters. They made them strong and *venenosic* and they became so subtle as to be marveled at.

When they made their matter thus subtle through sublimation and hot on account of the *corrosive* matters, and they had drawn enough of the tinctures into themselves, they *incerited* it with well cleansed ☿, that is, they poured a large quantity of ☿ over it, put it in tripodem, let it rise up and down till *Mercury* stayed with it. Thus they made their spirits fusible according to their will. They also took *calcined* ☽ and ☉, made into a most subtle *calx*, poured purified ☿ over it, set it *in tripodem* in a glass as is illustrated below, turned it over often and let it stand thus, *sublimating*, till the ☿ stayed with it. In this way they made the calx fusible. They *tinged* with it, that is, they dissolved the calc ⁘✚ , converted it into subtle crystals, which they cleansed well. Then they pulverized it, poured fresh ☿ over it, and set it *in tripodem* as has been discussed about ☉ and ☽. And thus they also made a medicine.

I am telling you dear *Filii*, the whole Art consists in the *ceration*. Therefore, read this over often, because it contains great wonders. You can make a *medicine* from all metals, as we have taught and spoken of, in a short time, without special work, harm, or cost, and all with this Art, with the *Mercurius Philosophorum*.

NOW I WILL TEACH HOW YOU CAN MAKE THE STONE FROM BAD WATER
AS WELL AS HOW TO MAKE THE *OLEA* OF METALS, WHICH CAN BE DONE WITH LITTLE WORK AND WITHOUT SEPARATION OF THE ELEMENTS AND TO BRING THEM TO SUCH A *PERFECTION* AS IS CERTAIN AND GOOD.

The reason why the elements are separated is that the *Imperfectum* be made *perfect*; also that the uncleanliness (or: impurities) be separated, and that the corpus and spirit be thus rid of all impurity, and be afterwards again conjoined. Know then, that anything that reaches the fire, no matter how impure it is, is made clean and pure by the fire, as we have taught before. The first sign is a perfect blackness, and we see it with our eyes. All matter becomes as black as pitch. Why? The fire drives *Corruption*, or what is rotten, upwards and it leaves the matter because of the strength of the heat. This is not done, however, with a strong but with a gentle fire. Then the *Corruption* or *feces* which are in the *Arca* are driven above until everything is black.

That is why *Morenius* says: Take care that you regulate your fire in such a way that you do not obtain whiteness before blackness. (the albedo before the nigredo) or all your work is spoiled by the whiteness if

283

it occurs before the blackness. So it must be a sure method that will drive the *Corruption* out by fire, and thus must be the purification and *Perfection*. Be careful in our work for after long and steady boiling the heat consumes the *Corruption*, feces and blackness, and changes it into another color, and ever another until it is perfectly white like snow. And it is done gently, so that the elements are not forced, but are gently rectified of their impurities. Take care, however, in every respect, as *Morenius* has warned, that you do not get the redness before the whiteness; for our stone must not be burnt in this work. Know that this is the best Way; for it is often necessary to give strong heat where the separation of the elements must be accomplished, before the *elementum ignis* is brought over, and everything must glow.

After this, if you wish to calcine the *feces* the matter has to be burnt in the reverberating furnace. Often the matter turns white; then it has to be changed into glass, and thus one thing is spoilt with the other. But, in the Great Work, there is no uncertainty. The *feces* know how to consume themselves of themselves, as *Geber* says: The dragon must devour its own blackness, and it has to be fed with its own venom. *Dantin* says: The black crow must hatch its own eggs with its young, till they all turn white. For that is the art and nature of all thing under the sky, that they desire to rectify themselves out of an inherent impulse and to rid themselves of their *feces* which are superfluous to them, and to be without defect For they were perfect and without defects from the beginning. The four *Elementa*, and everything made from them, mobile and immobile, nothing excepted, are all perfect in the beginning and in the end, and all things desire to be rid of their *feces*.

Someone might ask: but what <u>are</u> the feces? It is a *humor* or humidity (or: moisture) which God has ordered, and everything under the course of heaven must be nourished by it. It keeps all things in its nature and is in all things a perfect, elementary, natural warmth or fire, and it is a consuming and combustible fire. When the two are mixed, and if the perfect fire does not meet with unfortunate accidents, it will keep the thing in its nature. But as soon as a bad accident happens to the fire, which is also hot and imperfect, and one thing mingles with another, they all become hot and burn and destroy the thing, be it in metals, animals, trees or herbs, and in all things under the sky.

There are two kinds of water in all things created out of the Elements, a natural one and an elementary one, and that is perfect, good and eternal. Then there is still another water. It is called "water of the clouds". That is *imperfect*, and is mixed with the elementary water. It is meant to give nourishment and moisture to things and to keep them in their nature as long as no other extraneous water is added to them. But if more is added, it will drown the thing, so that it dies and corrupts, just as when water is poured into fire.

Similarly, you must understand this in regard to air and earth. If there were no *feces* in the elements, all things would be perfect, spiritual, and subtle, as God had meant them to be. Nor would there be decaying and death, as is explained about *feces* and diseases of the elements in the *Vegetabili*. Find it in Chapter 16.

Now you might ask, however: if a thing is destroyed in such a way, where then is the *perfection* which it contains? Read about that in the Vegetabili, Chapter 29. You will also find explained there how one thing attracts its like. Know also that if a thing has died, be it sensitive or insensitive, the spirit of its *corpus* separates from it and joins its like, from which it has originated, as you will understand by the *Vegetabili*. Look at the flame of fire or coal: the flames heat, the smoke moves upward. In this smoke is hidden the spirit of air. It joins its like. The same applies to the other elements.

But now someone may ask: Where then do the *feces elementorum* stay, when each thing has gone to its own? Concerning this, consider this example: If you put a glass vessel containing water into the sun, the sun draws the water to itself, and stinking black dregs (matter) will stay behind. Let it stand in the glass protected from rain and wind, for a long time, and the slimy black matter will in time become as white as snow and its smell will disappear. Such is the effect of the nature of the sun. Another example: Take a glass basin full of green herbs; put it in the sun or exposed to the air. The herbs will begin to decay and smell bad, and, each element draws towards its like as mentioned before. The black stinking earth stays in the basin, but after a long time the air and the warmth of the sun will calcine it as white as snow. And this is the work of nature.

Another example: Take the corpse of an evildoer, who lies on the rack or hangs on the gallows. The air and the sun consume its stench and decay, so that nothing remains but white ashes. In time the hard legs, which were full of fat and marrow, are thus consumed, so that they turn into a white, fine sand (salt) which is intangible between the fingers. That is brought about by nature, as we may see every day with our own eyes. Where then remains the stinking matter? It passes away and turns into nothing, and the element earth is thus cleansed and white as snow, so that it becomes impalpable. Thus it is evident in our Art one must not separate any elements, nor does one require and washing or purification concerning that it has to be tested to ascertain that it is good and penetrating. I have related to you so that you should understand that the *separationes elementorum* are not necessary in our work. Neither is *rectification*, because the *feces* consume themselves, as indicated; but in the *separationes elementorum*, a little is always lost in the fire, for they stand in the fire. And just as it easy to lose something, so it is to the detriment of the work, which you need not be afraid of in this instance, because in the Great Work no element is separated.

In addition, you should also know that one can make oil from all metals also without separation of the elements, and without much washing and dissolving. Yet, it must be done with *Aqua Fortis*, and you must give it a *ferment*, if you wish to make them from a perfect metal. But I advise you not to make oil from any imperfect

metal, except from ♄ and ♃ , one for the Red, the other for the White. There exist many different matters, from which to make oil. They have been discovered due to the rapidity in which they can be made, while some did not have the patience to endure the long amount of time required to accomplish the Great Work and because they seek small gain. Yet in such things there is great danger, more than in the Great Work, also greater labor and handicraft. You must distil *AF*, and you must also sublimate and be well acquainted with many unusual types of work. It also requires a great deal of money, effort and cost.

THE OTHER WORKS OF THE ANCIENT PHILOSOPHERS
SOME OLEA EX AQUIS FORTIBUS & METALLIS
Oil from with Brandy (or: Whisky)

There were some who made an *Aqua Fort* from Vitriol and ☉. In it they dissolved fine ☽ one part, of the cupel. After that, they ground it and washed the calx off with common water; then they dried it at the sun or with fire. Afterward, they put this calx into two glasses, poured rectified vinegar upon it, each time one lb. of vinegar upon 1 ounce of *calx lunae*. They put one of the glasses in the *balneum* and the other in front of it. Then they distilled the vinegar from one *Luna* onto the other, alternating the glasses; one <u>into</u> the balneum, the other one, which had been standing before it, out. Then they distilled once more, and did this until the ☽ was fully dissolved.

When the luna IS totally dissolved, the *Aqua Vitae* has to be drawn off in the *balneum* with gentle heat, such that one can suffer one's hand to be in this heat. When a skin forms on the Luna, the process must be stopped, allow it to cool down and put into a cold cellar to crystallize. Take out the crystals formed and put them in a small retort, lute it well and set it in warm ashes *in tripodem*, or let it stand until the clear little stones have been transformed into oil and no longer coagulate. This oil is a perfect elixir to make projection with it *ad album*.

ANOTHER KIND OF OIL MADE FROM *AF* AND LUNA

There have also been others who took 1 ounce of ☽ *amalgamated* with prepared ☿, in such a way that the *amalgamation* could be pressed through a linen cloth. After that, they set it for six weeks in a moderate heat; and they dissolved it in an *AF* made of Vitriol and ☉; drew it off again gently in the *balneum*. Then they removed it, stoppered it well, set it *in cinerum* or *tripodem* and gave it heat as if one wished to keep ♄ In flux. They kept it thus until the oil was fixed, and tested it in the following way: They took a sheet of and heated it to glowing, then poured one drop of this oil upon it. If the oil goes through without smoke, like oil through leather, and if it tinges to ☽, it is fixed, good and a perfect elixir. But if it does not do that, put it back *in tripodem* until it is fixed and transmutes ♀ , ♃ and ☿ into true ☽, which passes all tests.

A PRECIOUS OIL TO THE RED

Dear Sons, you should know the following and consider it a great secret. Take 3 lbs. of *vitriol virid. aer.*, *plumbum album an. 5 lbs croci pulv. lapidis haematitis* ℥ 4. *Saltpeter ad pondus omnium.* Crush them well so that they mingle thoroughly; divide them into three parts. From one of the parts, make an AF (aqua fort) in a glass vessel and no other kind of vessel. After this, pour it on the other part of the matter (a second part) and draw it over on a strong fire. Pour this now onto the third part of the matter and keep this water well closed.

Pulverize the Death's Heads and rub them with ✶▽, which I will teach later on, on a marble, till they are quite small, as if one wished to paint with the matter. Let it dry in a room or by the sun; grind it once again

290

and put it into an alembic. Pour your water on it, draw it off again, first with a gentle fire for 24 hours; then gradually with stronger fire, till the matter begins to glow. After this, keep it in a steady glow for 6 weeks. Then let it cool down, remove it and preserve it.

After this, take the Death's Head and the remaining feces. Powder them and moisten with ,vinegar and draw off its salt as you know how to do, so that no feces stay behind. When your salt is clear, pour the AF on it, give it gentle heat on sand or on ashes for 12 hours. Follow this with stronger heat for 6 hours, so that it will glow mildly. Then let it cool off. Take the water in the *recipient* and close it well. Again rub the feces with vinegar, and afterwards dissolve it in vinegar; put it in the *balneum*. Do as you have been instructed before and see if it produces feces. Coagulate it and pour the A.F. back on it. Draw it off. Repeat 3 or 4 times and the salt together with the water will go over the helm.

Do believe me that I have worked wonder with this water, which cannot be described here. I have personally turned this water into a red crystal which gave off a light at night by which a whole table of people could see enough to eat their meal by. Keep it until you need it, and consider it a treasure of all waters. More so take

Merc. praeparati of its humidity, for each pound of ☿ , 2 pounds of Vitrioli Romani and sublime it there through. Mix it again with the feces and for the third time, take fresh vitriol and sublime it again, Do this 4 or 5 times, the more times the better. After this, the ☿ is ready.

Take then, one ounce of ☉, thinly beaten and cut into
rolls. Dissolve it into the AF which you have made, and
set it into a basin with sifted ashes. Put the basin into
a 'kettle filled with very warm water, and in an hour the
will (symbol missing) will dissolve. (or sooner)

Take one ounce of the sublimated ☿; dissolve it also
in this water. After this, throw an other ounce of your
☿ in and let this also dissolve. Then it is enough as
you will have three ounces (of matter) dissolved in it, 1
oz. of ☉ and 2 oz. of ☿. Now put a helm on together
with a *recipient*, draw the water off, pour it back on
again (repeat) until it will now longer go over *in
balneo*. Let it cool down and put it in a furnace and
ashes. Lute a recipient to it, distill it over and pour
it back on again. Continue this so long as the water will
go over. In this distillation you will see wonders,
because you will see all the colors of the whole world in
the helm. The colors are in the spirit and the corpus
keeps the spirit in it and with it. The colors are
covered in the *corpus* as you will learn in VEGETABILIE.
Search for it in Chapter 93.

When no more drops are coming, let it cool down,
remove the helm and close the glass well above Set it *in
tripode* for 40 days; the heat should be such that you can
easily keep your hand in the furnace. Your matter will

292

become fixed within this time, and when it is cold, it will be hard as glass. As soon as it gets near heat that will melt wax, it flows as if it were wax or as an oil. This is a perfect *Lapis Compositus*, and no foreign things have been added to it which are not of its kind or species. My child should note that at least one part of this Stone falls on 1000 parts or more. I myself have worked in this area and have accomplished the operation one time. It is such a beautiful Stone to behold and shines so much at night that one does not have a need for light. This is why it is such an excellent *Medicine* and a noble Stone and should be considered a great *Secretum*.

The Water *Salis Arm.*, with which the above mentioned powder is to be rubbed or ground, is made in the

following way: Take 1 pound of ✳; 2 pounds *Vitriol* and sublimate them together and again mix the matter with the feces. For the third time, take fresh *Vitriol* and sublimate this also four times. Grind this sublimated

✳ to powder, put it into a glass, pour distilled vinegar upon it, just enough vinegar to dissolve it and no more. Now the water is as yellow as *Sol* when it has been sublimated through *Vitriol*. The *Vitriol aceturn destillatum* produces the *Tincture of Sol*. This then is the water which you must rub (grind) your Deaths Head, as

indicated above, which is to be *imbibed* with this ✳

▽. It gives good *Ingressum*.

THE SALTS OF THE METALS

Up to now, dear Son, you have heard how you are to proceed with and handle the Great Work and *Amalgamations* and with certain *Olea*. Now you will hear how to make salts out of the metals that can also produce a perfect *Elixir*, as good as the *Olea*, although its *projection* is not as high. It is an easy work, however and takes but a short time. After that, I will teach you how to make the Stone, which I consider my greatest *Secretum*.

Know that Salt can be derived from all metals and all salts of metals are *Elixirs*. They are the *Elementum terrae* under which the fire is hidden, because in the metals there are four elements, such as *Ignis*, *Aer*, *Aqua* and *Terra*. Fire and Earth are the outer elements, water and air the middle. The two outer are fixed, but the two middle elements (the inner) are volatile as water & air. You must know, however, that the *Elementum Ignis* can be separated from earth in all things combustible. It is its oil, except it cannot be done with metals, for they stay fixed together in the fire. That is why all Salts of Metals are elixirs.

Note further that after the *Salia Metallorum* have been turned into Elixir, its Projection is small. But when the same Salt is put together with an oil and its innermost part is brought out, and its outermost part is brought inside, where previously one part fell on a 100 parts, it will now fall upon 1000 parts; and as you projected it before on ☽ similarly you can transmute it afterwards on ☉ . When *Elixiria* are made from Salt, one can easily produce oils, as will be taught at the end of this work.

You must know that there exists no surer nor shorter way for working with the *Salia* of metals, for herein one cannot fail, nor can any infirmity befall you. Reason: there are no *spiritus* that could evanesce. Also, it is difficult to do things wrong with the fire, because the matter is not congealed, as it is fixed already, and does not require your effort to do so.

The old *Philosophi* swore to each other that they would not reveal the two secrets, how to make Salt and Oil out of metals, to anyone except their children, who would be ready for it. They also wrote of it in such a way, that no one could understand it except the Children of the Art. Read all the books of the ancient philosophers, and you will nowhere find correct information, neither on the Salt nor on the Oil of metals, which would enable you to make them of these substances. They may well write that the *Salia Metallorum* are *Elixiria*, and say that all *Philosophi* agree in this, and that it may be easy to achieve through the reverberation of the metal. They may also say you should draw it from its *Fecibus* and work with it until it is a crystal-stone. Yet, they do not write anything else about this work, or write in such obscure terms that no one can understand anything. Thus the arts of salts and oils has remained hidden more than all the other arts. I am telling you truthfully that the art *extrationis salium metallorum* has never been revealed to anyone in my time.

I therefore entreat you by the living God, that you do not reveal this secret to anyone but those of whom you are sure and certain that they will keep it secret on account of the many evils it might involve and because the noble Art might be consumed and used in sin; and God's honor and praise and the poor needy might be forgotten. Therefore, note carefully, to whomsoever God gives the art, he has it and no one else; according to whether his intention is good or evil, God bestows it. Enough said to those who have understanding.

Take then ☉ and ☽ , dissolve them in AF and beat it to the ground. Wash the *calcem* with *Aqua Dulci* and dry it. Then it is ready to be put in the calcination furnace, to open it up, so that the ☿ can be sublimated out of it. This not only applies to ☉ and ☽ but also to the other metals such as ♄, ♃, ♀, ♂ & etc.

Yet those from ☉ and ☽ are the best; they also make a higher projection and are easiest to process to oil and elixir. Now take the *calx*, put it into a glass vessel with a wide bottom. Put the *calx lunae* in it, one finger's thickness, neither thicker nor thinner. Set it thus into the calcination—furnace where the spirits are calcined within, or into the *Athanor* or *in Tripodem*. Heat it, as if one were to keen ♄ in flux without driving it, for 21 days. Do not let the fire go down, so that the *Corpus* may open up and let ☿ go. Know that you must thus proceed with all metals, be it ☉ or ☽ or other metals. Only, ☉ must stand for six weeks, because it is a *Compact* and a perfect *Corpus*, which ☽ and the other metals are not. That is also the reason why they must not stand as long in the reverberating furnace.

When now your *Calx* of the metals has thus been opened, take an alembic with a wide bottom, put your calx into it at equal thickness. Lute a big helm on it, put it on the sublimating furnace in ashes, lute a receiver on it, light a fire under it, at first, small, then increase it by degree until it begins to glow. Keep it glowing thus for 8 to 10 days, watch to see if some ☿ continues to sublimate. If nothing comes out, let it cool down, take the helm off; you will find Mercurius sublimated (☿), white as snow. Save.

You will find the salt or the earth at the bottom of the glass, in the form of a greyish—white powder. It is swollen like a sponge, while Mercurius has been leaving it. It is the same kind of process with ☽ , ♀ and ♄ , except that the *Calx* must stand for 16 to 20 days, and must also glow stronger, before ☿ comes out of it; for ☉ does not melt as easily as the other metals, because it is much firmer than ☽ or any of the other metals.

Now remove the salt or earth from the glass, put it into a stone jar and pour distilled wine vinegar on it, set it into a boiling *Balneum*, let it stand from 4 to 6 days, stirring frequently, it should be well covered. Then allow it to stand and cool down, pour it off and preserve it. Upon the *Feces* pour fresh *acetum distillatum*, and set it into the *Balneum* for 24 hours, stirring frequently. Let it stand, cooling, and pour it together with the other. If you feel there is still something left in the *Feces* pour more vinegar upon them: if not, throw them away as they have no further use.

Draw the decanted vinegar off in the *Balneum*, and the salt will remain at the bottom clear and snow white. Following this, pour clear, pure water on it, set it back into the balneum, let it dissolve so that *Feces* appear. Throw those away, coagulate the Salt, and dissolve it again with *Aqua Communi*. Repeat this process until no more feces appear. Now it is ready to make the first *Projection* on ☿ with it. To extract the salt (or: to draw the salt) from ☉ and other metals it is an easier process than the one related here.

Further, if you wish to get the salt from imperfect metals, from copper and iron, you must also know that they need to be filed finely, set them into the reverberating furnace for 6 days in a moderate heat and glow. After that, you can draw out the salt as we have taught about ☽.

If you wish to get salt from ♄ and ♃, you must dissolve them in *Aqua Fort*, draw off the *Calx* and reverberate it, as has been indicated. However, if you wish to get only the salt from ♄ and ♃, without *Mercurius*, let it stand in a reverberating furnace for 12 days in a rather strong glow, but such that it does not turn into a glass.

You should not put on more than one finger's thickness of Calx, it will then swell up like a sponge. After this, extract the Salt, as you have been previously instructed. This Salt is as good as the Salt of ☽ , and achieves a high projection. One projects it upon *Mercury*, and the salt *plumbi* makes as high a projection on *Mercury* as does the Salt derived from *Sol*.

If, however, you wish to retain *Mercury plumbi* and ♃, sublimate it out of it, as is done with ☽, except that you must not reverberate for 20 days in the *Athanor* because its ☿ is not fixed in *corpore*. Then it separates, breaking away from the *corpus*. This is the

299

best way to draw out the ☿ and to sublimate it, for thus
each element retains its power. ☿ and ♀ must
reverberate a long time, ♂ 98 days, ♀ 35 days. These
two must reverberate for such lengths of time because of
their *feces*. They must be annealed slowly, to prevent
their turning into glass; because they do not easily
their *Mercury*.

When they are well opened, extract the Salt with the
vinegar. When nothing further comes out, set the matter
to reverberate again for three days, till nothing draws
out. When the Salt has been processed, make projection
upon *Mercury*, because Salts transmute (or: they
transmute) all the *Mercury*, while they have little or no
spirit in them.

Take *Mercury*, let it get as hot as possible, and
throw it upon the salt of metals, 1 part to 100 parts;
increase the fire so that it flows strongly, as ☽ in
(or: on) the cupel. Let it stand thus until it settle
into a King. Then slake it immediately as is required.
Now you have fine ☽ . All the Salts of metals, be they
red or white, only produce ☽ ; but perfect metals make a
higher projection than imperfect ones. After they have
been reversed, however, and their innermost turned
without, they change into oil. Then they all make
projection to the Red, and where before they made 1 into

100 parts, they now make projection 1 to 1000 parts. Before, one could only throw upon *Mercury*, but when they have become oils, they make projection upon all metals, as will be taught herein.

OILS FROM METALS

Now I will teach you how to make oils from metals, and to turn the innermost outside, which is one of the greatest secrets, for after that process they will make ☉, while before they only could make ☽. Now we will proceed with the projection.

Take a large amount of *Vitriolum Romanum* (probably copper sulphate — hwn), 12 or 16 lbs., more or less. Dissolve it in *Aqua Communa*, and when it is dissolved, let it settle down. Pour off the clear from its *feces*, and set it into a sandbath to coagulate, using a good vessel. Let it evaporate until a skin forms atop. Now let it cool down. In this way the Vitriol sprouts into a ~iful green, which is the best color in the *Simplice*.
~ sprouted into a suitable vessel. Let the
~orate and sprout (or: shoot) until you
· put it into a room for it to
heat as the sun gives off in
.ll turn white.

ore earth will drop to the
vaporate and sprout again, as
. This can be accomplished

for
ng it,
ght that
be
outside.

301

within 3—4 days. The Vitriol will become twice as beautiful as before, and much greener, so much so that you will not have seen a more beautiful green color. Therefore *Hermes* and *Geber* speak: Preserve well your green, evaporate well the wet until a skin forms on top; then let it sprout again, and continue to do this till you have your Vitriol together again. Put it once more in a room, as before, drying until it turns white. Then dissolve and *granulate* again. Repeat until no further *feces* are left. Put aside until you need it.

Now then, someone might say: When the Vitriol has been dissolved, why do you not let it evaporate completely, but you allow it to sprout, and it requires a great amount of time to change it into a white powder?

Note, then, that Vitriol has within it, a subtle *spiritus*, as is described in the Vegatabile Work as regards the subtle spirits, which are in all herbs outside. It is the green (or: verdure) of all things that are green outside, for it is the flower of their right *essence*. If you lose the greenness, I am telling you forsooth, that you are deprived of the *essence*. Further, whatever work you are doing with it, it is all lost, in vain, for it has been deprived of much, its soul, life and *essentia*, *corpuse*, *spiritus* or roots and everythin green outside. Take care then, to preserve that well it is so subtle that you can lose it without notic as indicated in the Vegatabili. There you are ta the green in all herbs, leaves and roots, is t extracted from everything that is green on t

302

And when the green has been drawn out and reversed into a beautiful redness, the like of which none has seen, you have the right *essentia*.

Look for further instruction in the Vegatabili. This is the reason why it is necessary to preserve the green of the Vitriol. If you were to coagulate it, part of its greenness would be taken from it; for it would become yellow, while yet the green is in the *Quintessentia* that we seek in the Vitriol. That is why you must permit it to dry and sprout in a room. Then its greenness is covered with the white, for as soon as it becomes moist again, its greenness will re-appear. Thus the outermost of the Vitriol must be turned into the innermost, and the innermost must come out, in order to preserve its soul and its spirit and to retain its *Quinta Essentia*. This is a great *Mysterium* or *Secretum* in our Art. When the Vitriol has thus been cleansed, it is as red as a rose or ruby. It has within itself the four elements in their perfection, and this is the stone which God has given us for nothing.

You should now take the white powder which you were told to put aside and place it in a phial and close it with *Sigillo Hermetis*. Set it in ashes and heat it by a lamp, as warm as the sun shines in the midst of summer. Keep it thus, until you see that it begins to turn yellow. Let it stand further until it turns completely yellow. Then, let it stand yet another ten days and see if it does not begin to tinge a red color. Then, increase the fire a little, and if it becomes more red, let it stand in the regimen of the fire as is. If however, it does not become somewhat redder in 8-10 days, increase the fire by one lamp until the color increases. If it stays the same, add yet another lamp, thus each time increasing the heat by degrees until the color changes to a rose or ruby red.

When it has become a high or deep red color, let it stand yet another 8—10 days in the same heat and watch if the color does not change into a color different than red. Now the matter has been reversed and its innermost has been brought outside. In this way, you will not lose the greenness if it has been reversed into redness. This is because it is in the deepest inner parts and can no more be brought out. It will forever stay red and unfixed, for if it were fixed, everything would be lost, because it would have to be dissolved in water and coagulated again, and afterwards distilled over the helm.

I am telling you that I have never revealed to you greater secrets than this! I am telling you, by my God, that this *SECRET* has never been set down into writing by the *philosophi* except by my hand alone. Moreover, I am telling you that there is no greater *secret* in art than this. Therefore, I beseech you and all those who will understand it, that you will never bring it to light except where it is right to reveal it, by the damnation of your soul, for it is a *Secret* above ALL *Secrets*, since with this matter all metals can be turned into oil, when they are dissolved in *Aqua Fort.*, when the *calx* has been beaten to the bottom and processed as required.

All *Olea Metallorum* turn red as blood, without ☽ and

❓ not (symbol missing) for all metals are red in their

innermost, but one is redder than the other. When they
have been brought to redness, you must dissolve them,
again coagulate them until they are free from all *feces*
and they have their elements perfectly joined (together);
for once they have arrived at this stage, nothing is left
but *feces*. The earth, too, has become subtle and liquid
and is dissolved in the other three.

When they have thus been made subtle, with dissolving
and coagulating, you can distil it over the helm to a red
oil, as you will learn As you are working with *VITRIOL*,

you must also treat ☿. After it has been dissolved in

Aqua Fort., beaten down, decanted from its saltiness and
dried, you put it in a glass the same as has been done
with the *Vitriol*. Or you can put a *sublimated Mercury*
into such a glass, proceed in the same way and cleanse it
of its *feces*, and distill it over into a red oil. In the

same way ♀ can be processed. What do you think? Is this

not a great *Secret*? Never before has anything like it
been heard. Open your ears therefore, listen AND
understand!

Now we will return to our work. When you see that
your matter remains in an oily state, take it out of the
ashes and put it into another, strong, glass. Pour a

goodly amount of wine vinegar upon it, and set it into the balneum to boil for 4 days, often stirring it with a wooden spoon. After the fourth day, let it cool down and settle. Decant off the clear liquid and pour more vinegar upon the remaining *feces*. Add more distilled vinegar, and repeat three times. Now throw away the *feces* and put an *alembic* upon the glass containing the *solution*; draw off the vinegar, so that the matter becomes quite dry. Now you have the matter at the bottom of the glass and much more beautiful than before. Again, pour fresh vinegar upon it, and treat it as above. Reiterate this until no more *feces* remain in the *Solution*. Then *coagulate* it to a dry powder, put a helm on with a large head (*caput*) and distill. First you will obtain a yellow *spiritus*, then red oils and finally a white *spiritus*. Let the matter cool down, remove the receiver and its contents. It is the blessed Oil. Preserve it well until you need it for your metallic salt.

At the bottom of the alembic you will find a matter that is as white as snow and as clear as crystal. It is the *rectified* matter of the aforesaid *materia*. It can be pulverized and imbibed into the red oil as into its own corpus. Put it in *vitreum apullam* and hang it *in tripodem* for 40 days in moderate heat. Now it will coagulate into a LAPIS PHILOSOPHORUM which will dissolve all metals into ⊙. But we will not do this now, but will work toward our Salt and oil of metals in this manner, as with *Vitriol*. Thus the element of earth will go over with the oil, red as blood. This the earth of *Vitriol* does not do, as its oil separates from the earth. Consequently God has given it such *Benediction* that from it alone can one make the *LAPIS PHILOSOPHORUM* without any *Addition*. But first one has to fix its oil with its earth. That does not happen in metals, because their earth goes over the helm together with the fire, and the whole body reverses, which tinges the metals into perfect (symbol missing).

By the same process, you can make the oil of ☿ and ♀, and the earth also goes over the helm in the oil and stays in the oil for all eternity. With this oil, you can perform such miracles as would be too lengthy to recount here. You well know what is said about the oil *Veneris*.

Yet the oil from ☿ is much better in its effects than the oil *Veneris*.

-end of the Mineral Work-

A Word from the Publisher

Thank you for purchasing this small work from The R.A.M.S. Library of Alchemy. During his lifetime, Hans Nintzel was dedicated to the identification, acquisition, study, retyping and, when necessary, translation of what he considered to be the most important known works on Alchemy. Hans was assisted by his sparse network of fellow Alchemists, all members of the Restorers of Alchemical Manuscripts Society (R.A.M.S.). I was an active member of R.A.M.S.

My goal is to publish all of the works originally made available through R.A.M.S. as photocopies. To facilitate this, I have chosen to have the books professionally printed. I also have a few titles that I intend to add to the original R.A.M.S. Library, selected by strict criteria established by Hans.

If you have a work on Alchemy that you believe should be a part of the R.A.M.S. Library, please contact me through R.A.M.S. Publishing Company.

Philip N. Wheeler